Computational Analysis of Randomness in Structural Mechanics

Structures and Infrastructures Series

ISSN 1747-7735

Book Series Editor:

Dan M. Frangopol
Professor of Civil Engineering and
Fazlur R. Khan Endowed Chair of Structural Engineering and Architecture
Department of Civil and Environmental Engineering
Center for Advanced Technology for Large Structural Systems (ATLSS Center)
Lehigh University
Bethlehem, PA, USA

Volume 3

Computational Analysis of Randomness in Structural Mechanics

Christian Bucher

Center of Mechanics and Structural Dynamics,
Vienna University of Technology, Vienna, Austria

CRC Press is an imprint of the
Taylor & Francis Group, an **informa** business

A BALKEMA BOOK

Colophon

Book Series Editor:
Dan M. Frangopol

Volume Author:
Christian Bucher

Cover illustration:
Joint probability density function of two Gaussian random variables conditional on a circular failure domain.

Taylor & Francis is an imprint of the Taylor & Francis Group, an informa business

Typeset by Macmillan Publishing Solutions, Chennai, India
Printed and bound in Great Britain by Antony Rowe (a CPI Group company), Chippenham, Wiltshire

British Library Cataloguing in Publication Data
A catalogue record for this book is available from the British Library

Library of Congress Cataloging-in-Publication Data

Bucher, Christian.
Computational analysis of randomness in structural mechanics / Christian Bucher.
p. cm. — (Structures and infrastructures series v. 3)
Includes bibliographical references and index.
ISBN 978-0-415-40354-2 (hardcover : alk. paper)—ISBN 978-0-203-87653-4 (e-book)
1. Structural analysis (Engineering)—Data processing. 2. Stochastic analysis. I. Title.
II. Series.

TA647.B93 2009
624.1'710285—dc22

2009007681

Published by: CRC Press/Balkema
P.O. Box 447, 2300 AK Leiden, The Netherlands
e-mail: Pub.NL@taylorandfrancis.com
www.crcpress.com – www.taylorandfrancis.co.uk – www.balkema.nl

ISBN13 978-0-415-40354-2(Hbk)
ISBN13 978-0-203-87653-4(eBook)
Structures and Infrastructures Series: ISSN 1747-7735
Volume 3

Table of Contents

Editorial

Welcome to the New Book Series *Structures and Infrastructures*.

Our knowledge to model, analyze, design, maintain, manage and predict the life-cycle performance of structures and infrastructures is continually growing. However, the complexity of these systems continues to increase and an integrated approach is necessary to understand the effect of technological, environmental, economical, social and political interactions on the life-cycle performance of engineering structures and infrastructures. In order to accomplish this, methods have to be developed to systematically analyze structure and infrastructure systems, and models have to be formulated for evaluating and comparing the risks and benefits associated with various alternatives. We must maximize the life-cycle benefits of these systems to serve the needs of our society by selecting the best balance of the safety, economy and sustainability requirements despite imperfect information and knowledge.

In recognition of the need for such methods and models, the aim of this Book Series is to present research, developments, and applications written by experts on the most advanced technologies for analyzing, predicting and optimizing the performance of structures and infrastructures such as buildings, bridges, dams, underground construction, offshore platforms, pipelines, naval vessels, ocean structures, nuclear power plants, and also airplanes, aerospace and automotive structures.

The scope of this Book Series covers the entire spectrum of structures and infrastructures. Thus it includes, but is not restricted to, mathematical modeling, computer and experimental methods, practical applications in the areas of assessment and evaluation, construction and design for durability, decision making, deterioration modeling and aging, failure analysis, field testing, structural health monitoring, financial planning, inspection and diagnostics, life-cycle analysis and prediction, loads, maintenance strategies, management systems, nondestructive testing, optimization of maintenance and management, specifications and codes, structural safety and reliability, system analysis, time-dependent performance, rehabilitation, repair, replacement, reliability and risk management, service life prediction, strengthening and whole life costing.

This Book Series is intended for an audience of researchers, practitioners, and students world-wide with a background in civil, aerospace, mechanical, marine and automotive engineering, as well as people working in infrastructure maintenance, monitoring, management and cost analysis of structures and infrastructures. Some volumes are monographs defining the current state of the art and/or practice in the field, and some are textbooks to be used in undergraduate (mostly seniors), graduate and postgraduate courses. This Book Series is affiliated to *Structure and Infrastructure Engineering* (http://www.informaworld.com/sie), an international peer-reviewed journal which is included in the Science Citation Index.

It is now up to you, authors, editors, and readers, to make *Structures and Infrastructures* a success.

Dan M. Frangopol
Book Series Editor

About the Book Series Editor

Dr. Dan M. Frangopol is the first holder of the Fazlur R. Khan Endowed Chair of Structural Engineering and Architecture at Lehigh University, Bethlehem, Pennsylvania, USA, and a Professor in the Department of Civil and Environmental Engineering at Lehigh University. He is also an Emeritus Professor of Civil Engineering at the University of Colorado at Boulder, USA, where he taught for more than two decades (1983–2006). Before joining the University of Colorado, he worked for four years (1979–1983) in structural design with A. Lipski Consulting Engineers in Brussels, Belgium. In 1976, he received his doctorate in Applied Sciences from the University of Liège, Belgium, and holds two honorary doctorates (Doctor Honoris Causa) from the Technical University of Civil Engineering in Bucharest, Romania, and the University of Liège, Belgium. He is a Fellow of the American Society of Civil Engineers (ASCE), American Concrete Institute (ACI), and International Association for Bridge and Structural Engineering (IABSE). He is also an Honorary Member of both the Romanian Academy of Technical Sciences and the Portuguese Association for Bridge Maintenance and Safety. He is the initiator and organizer of the Fazlur R. Khan Lecture Series (www.lehigh.edu/frkseries) at Lehigh University.

Dan Frangopol is an experienced researcher and consultant to industry and government agencies, both nationally and abroad. His main areas of expertise are structural reliability, structural optimization, bridge engineering, and life-cycle analysis, design, maintenance, monitoring, and management of structures and infrastructures. He is the Founding President of the International Association for Bridge Maintenance and Safety (IABMAS, www.iabmas.org) and of the International Association for Life-Cycle Civil Engineering (IALCCE, www.ialcce.org), and Past Director of the Consortium on Advanced Life-Cycle Engineering for Sustainable Civil Environments (COALESCE). He is also the Chair of the Executive Board of the International Association for Structural Safety and Reliability (IASSAR, www.columbia.edu/cu/civileng/iassar) and the Vice-President of the International Society for Health Monitoring of Intelligent Infrastructures (ISHMII, www.ishmii.org). Dan Frangopol is the recipient of several prestigious awards including the 2008 IALCCE Senior Award, the 2007 ASCE Ernest Howard Award, the 2006 IABSE OPAC Award, the 2006 Elsevier Munro Prize, the

2006 T. Y. Lin Medal, the 2005 ASCE Nathan M. Newmark Medal, the 2004 Kajima Research Award, the 2003 ASCE Moisseiff Award, the 2002 JSPS Fellowship Award for Research in Japan, the 2001 ASCE J. James R. Croes Medal, the 2001 IASSAR Research Prize, the 1998 and 2004 ASCE State-of-the-Art of Civil Engineering Award, and the 1996 Distinguished Probabilistic Methods Educator Award of the Society of Automotive Engineers (SAE).

Dan Frangopol is the Founding Editor-in-Chief of *Structure and Infrastructure Engineering* (Taylor & Francis, www.informaworld.com/sie) an international peer-reviewed journal, which is included in the Science Citation Index. This journal is dedicated to recent advances in maintenance, management, and life-cycle performance of a wide range of structures and infrastructures. He is the author or co-author of over 400 refereed publications, and co-author, editor or co-editor of more than 20 books published by ASCE, Balkema, CIMNE, CRC Press, Elsevier, McGraw-Hill, Taylor & Francis, and Thomas Telford and an editorial board member of several international journals. Additionally, he has chaired and organized several national and international structural engineering conferences and workshops. Dan Frangopol has supervised over 70 Ph.D. and M.Sc. students. Many of his former students are professors at major universities in the United States, Asia, Europe, and South America, and several are prominent in professional practice and research laboratories.

For additional information on Dan M. Frangopol's activities, please visit www.lehigh.edu/~dmf206/

Foreword

Computational Analysis of Randomness in Structural Mechanics aims at detailing the computational aspects of stochastic analysis within the field of structural mechanics. This book is an excellent guide to the numerical analysis of random phenomena.

Chapter 1 describes the organization of the book's contents and presents a collection of simple examples dealing with the quantification of stochastic uncertainty in structural analysis. Chapter 2 develops a background in probability and statistical concepts. Chapter 3 introduces basic techniques for regression and response surfaces. Chapter 4 describes random processes in both time and frequency domains, presents methods to compute the response statistics in stationary and non-stationary situations discusses Markov process and Monte Carlo simulation, and concludes with a section on stochastic stability. Chapter 5 deals with response analysis of spatially random structures by describing random fields and implementation of discrete models in the context of finite element analysis. Finally, Chapter 6 presents a representative selection of methods aiming at providing better computational tools for reliability analysis.

The Book Series Editor would like to express his appreciation to the Author. It is his hope that this third volume in the *Structures and Infrastructures* Book Series will generate a lot of interest in the numerical analysis of random phenomena with emphasis on structural mechanics.

Dan M. Frangopol
Book Series Editor
Bethlehem, Pennsylvania
January 20, 2009

Preface

As many phenomena encountered in engineering cannot be captured precisely in terms of suitable models and associated characteristic parameters, it has become a long-standing practice to treat these phenomena as being random in nature. While this may actually not be quite correct (in the sense that the underlying physical processes might be very complex—even chaotic—but essentially deterministic), the application of probability theory and statistics to these phenomena, in many cases, leads to the correct engineering decisions.

It may be postulated that a description of *how* these phenomena occur is essentially more important to engineers than *why* they occur. Taking a quote from Toni Morrison's "The Bluest Eye" (admittedly, slightly out of context), one might say:

But since why is difficult to handle, one must take refuge in how.[1]

This book comprises lectures and course material put together over a span of about 20 years, covering tenures in Structural Mechanics at the University of Innsbruck, Bauhaus-University Weimar, and Vienna University of Technology. While there is a substantial body of excellent literature on the fascinating topic of the modelling and analysis of random phenomena in the engineering sciences, an additional volume on "how to actually do it" may help to facilitate the cognitive process in students and practitioners alike.

The book aims at detailing the computational aspects of stochastic analysis within the field of structural mechanics. The audience is required to already have acquired some background knowledge in probability theory/statistics as well as structural mechanics. It is expected that the book will be suitable for graduate students at the master and doctoral levels and for structural analysts wishing to explore the potential benefits of stochastic analysis. Also, the book should provide researchers and decision makers in the area of structural and infrastructure systems with the required probabilistic background as needed for strategic developments in construction, inspection, and maintenance.

In this sense I hope that the material presented will be able to convey the message that even the most complicated things can be dealt with by tackling them step by step.

Vienna, December 2008
Christian Bucher

[1] Toni Morrison, The Bluest Eye, Plume, New York, 1994, p 6.

About the Author

Christian Bucher is Professor of Structural Mechanics at Vienna University of Technology in Austria since 2007. He received his Ph.D. in Civil Engineering from the University of Innsbruck, Austria in 1986, where he also obtained his "venia docendi" for Engineering Mechanics in 1989.

He was recipient of the post-doctoral Erwin Schrödinger Fellowship in 1987/88. He received the IASSAR Junior Research Prize in 1993 and the European Academic Software Award in 1994. In 2003, he was Charles E. Schmidt Distinguished Visiting Professor at Florida Atlantic University, Boca Raton, Florida. He has also been visiting professor at the Polish Academy of Science, Warsaw, Poland, the University of Tokyo, Japan, the University of Colorado at Boulder, Boulder, Colorado, and the University of Waterloo, Ontario.

Prior to moving to Vienna, he held the positions of Professor and Director of the Institute of Structural Mechanics at Bauhaus-University Weimar, Germany, for more than twelve years. At Weimar, he was chairman of the Collaborative Research Center SFB 524 on Revitalization of Buildings.

He is author and co-author of more than 160 technical papers. His research activities are in the area of stochastic structural mechanics with a strong emphasis on dynamic problems. He serves on various technical committees and on the editorial board of several scientific journals in the area of stochastic mechanics. He advised more than 15 doctoral dissertations in the fields of stochastic mechanics and structural dynamics.

He has been teaching classes on engineering mechanics, finite elements, structural dynamics, reliability analysis as well as stochastic finite elements and random vibrations. He was principal teacher in the M.Sc. programs "Advanced mechanics of materials and structures" and "Natural hazard mitigation in structural engineering" at Bauhaus-University Weimar. In Vienna, he is deputy chair of the Doctoral Programme W1219-N22 "Water Resource Systems".

He is co-founder of the software and consulting firm DYNARDO based in Weimar and Vienna. Within this firm, he engages in consulting and development regarding the application of stochastic analysis in the context of industrial requirements.

Chapter 1

Introduction

ABSTRACT: This chapter first describes the organization of the book's contents. Then it presents a collection of simple examples demonstrating the foundation of the book in structural mechanics. All of the simple problems deal with the question of quantifying stochastic uncertainty in structural analysis. These problems include static analysis, linear buckling analysis and dynamic analysis.

1.1 Outline

The introductory section starts with a motivating example demonstrating various random effects within the context of a simple structural analysis model. Subsequently, fundamental concepts from continuum mechanics are briefly reviewed and put into the perspective of modern numerical tools such as the finite element method.

A chapter on probability theory, specifically on probabilistic models for structural analysis, follows. This chapter 2 deals with the models for single random variables and random vectors. That includes joint probability density models with prescribed correlation. A discussion of elementary statistical methods – in particular estimation procedures – complements the treatment.

Dependencies of computed response statistics on the input random variables can be represented in terms of regression models. These models can then be utilized to reduce the number of variables involved and, moreover, to replace the – possibly very complicated – input-output-relations in terms of simple mathematical functions. Chapter 3 is devoted to the application of regression and response surface methods in the context of stochastic structural analysis.

In Chapter 4, dynamic effects are treated in conjunction with excitation of structures by random processes. After a section on the description of random processes in the time and frequency domains, emphasis is put on the quantitative analysis of the random structural response. This includes first and second moment analysis in the time and frequency domains.

Chapter 5 on the analysis of spatially random structures starts with a discussion of random field models. In view of the numerical tools to be used, emphasis is put on efficient discrete representation and dimensional reduction. The implementation within the stochastic finite element method is then discussed.

The final chapter 6 is devoted to estimation of small probabilities which are typically found in structural reliability problems. This includes static and dynamic problems as well as linear and nonlinear structural models. In dynamics, the quantification of first passage probabilities over response thresholds plays an important role. Priority is given to Monte-Carlo based methods such as importance sampling. Analytical approximations are discussed nonetheless.

Throughout the book, the presented concepts are illustrated by means of numerical examples. The solution procedure is given in detail, and is based on two freely available software packages. One is a symbolic maths package called `maxima` (Maxima 2008) which in this book is mostly employed for integrations and linear algebra operations. And the other software tool is a numerical package called `octave` (Eaton 2008) which is suitable for a large range of analyses including random number generation and statistics. Both packages have commercial equivalents which, of course, may be applied in a similar fashion.

Readers who want to expand their view on the topic of stochastic analysis are encouraged to refer to the rich literature available. Here only a few selected monographs are mentioned. An excellent reference on probability theory is Papoulis (1984). Response surface models are treated in Myers and Montgomery (2002). For the modeling and numerical analysis of random fields as well as stochastic finite elements it is referred to VanMarcke (1983) and Ghanem and Spanos (1991). Random vibrations are treated extensively in Lin (1976), Lin and Cai (1995), and Roberts and Spanos (2003). Many topics of structural reliability are covered in Madsen, Krenk, and Lind (1986) as well as Ditlevsen and Madsen (2007).

1.2 Introductory examples

1.2.1 *Outline of analysis*

The basic principles of stochastic structural analysis are fairly common across different fields of application and can be summarized as follows:

- Analyze the physical phenomenon
- Formulate an appropriate mathematical model
- Understand the solution process
- Randomize model parameters and input variables
- Solve the model equations taking into account randomness
- Apply statistical methods

In many cases, the solution of the model equations, including randomness, is based on a repeated deterministic solution on a sample basis. This is usually called a Monte-Carlo-based process. Typically, this type of solution is readily implemented but computationally expensive. Nevertheless, it is used for illustrative purposes in the subsequent examples. These examples intentionally discuss both the modeling as well as the solution process starting from fundamental equations in structural mechanics leading to the mathematical algorithm that carries out the numerical treatment of randomness.

1.2.2 *Static analysis*

Consider a cantilever beam with constant bending stiffness EI, span length L subjected to a concentrated load F located the end of the beam.

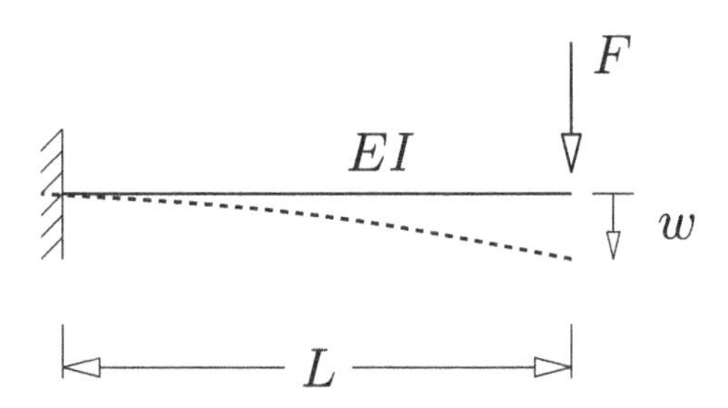

Figure 1.1 Cantilever under static transversal load.

First, we want to compute the end deflection of the beam. The differential equation for the bending of an Euler-Bernoulli beam is:

$$EIw'''' = 0 \rightarrow w''' = C_1; \quad w'' = C_1 x + C_2; \quad w' = C_1 \frac{x^2}{2} + C_2 x + C_3$$

$$w = C_1 \frac{x^3}{6} + C_2 \frac{x^2}{2} + C_3 x + C_4 \tag{1.1}$$

Using the appropriate boundary conditions we obtain for the deflection and the slope of the deflection

$$w(0) = 0 = C_4; \quad w'(0) = 0 = C_3; \tag{1.2}$$

and for the bending moment M and the shear force V

$$\begin{aligned} M(L) &= 0 = -EIw''(L) = -EI(C_1 L + C_2) \\ V(L) &= F = -EIw'''(L) = -EIC_1 \end{aligned} \tag{1.3}$$

So we get

$$C_1 = -\frac{F}{EI}; \quad C_2 = -C_1 L = \frac{FL}{EI} \tag{1.4}$$

and from that

$$w(x) = -\frac{F}{EI}\frac{x^3}{6} + \frac{FL}{EI}\frac{x^2}{2} \tag{1.5}$$

The vertical end deflection under the load is then given by

$$w = \frac{FL^3}{3EI} \tag{1.6}$$

Assume $L = 1$ and that the load F and the bending stiffness EI are random variables with mean values of 1 and standard deviations of 0.1. What is the mean value and the standard deviation of w?

One can attempt to compute the mean value of w by inserting the mean values of F and EI into the above equation. This results in $\bar{w} = \frac{1}{3}$. Alternately, we might try to solve the problem by Monte-Carlo simulation, i.e. by generating random numbers representing samples for F and EI, compute the deflection for each sample and estimate the statistics of w from those values.

This octave script does just that.

```
1  M=1000000;
2  F=1+.1*randn(M,1);
3  EI=1+.1*randn(M,1);
4  w=F./EI/3.;
5  wm=mean(w)
6  ws=std(w)
7  cov=ws/wm
```

Running this script three times, we obtain (note that the results do slightly differ in each run)

```
1   wm=0.33676
2   ws=0.048502
3   cov=0.14403
4
5   wm=0.33673
6   ws=0.048488
7   cov=0.14399
8
9   wm=0.33679
10  ws=0.048569
11  cov=0.14421
```

In these results, wm denotes the mean value, ws the standard deviation, and cov the coefficient of variation (the standard deviation divided by the mean). It can be seen that the mean value is somewhat larger than $\frac{1}{3}$. Also, the coefficient of variation of the deflection is considerably larger than the coefficient of variation of either F or EI.

Exercise 1.1 (Static Deflection)

Consider a cantilever beam as discussed in the example above, but now with a varying bending stiffness $EI(x) = \frac{EI_0}{1-\frac{x}{2L}}$. Repeat the deflection analysis like shown in the example

a) for deterministic values of F, L and EI_0

b) for random values of F, L and EI_0. Assume that these variables have a mean value of 1 and a standard deviation of 0.05. Compute the mean value and the standard deviation of the end deflection using Monte Carlo simulation.

Solution: The deterministic end deflection is $w_d = \frac{5FL^3}{12EI_0}$. A Monte Carlo simulation with 1000000 samples yields a mean value of wm $= 0.421$ and a standard deviation of ws $= 0.070$.

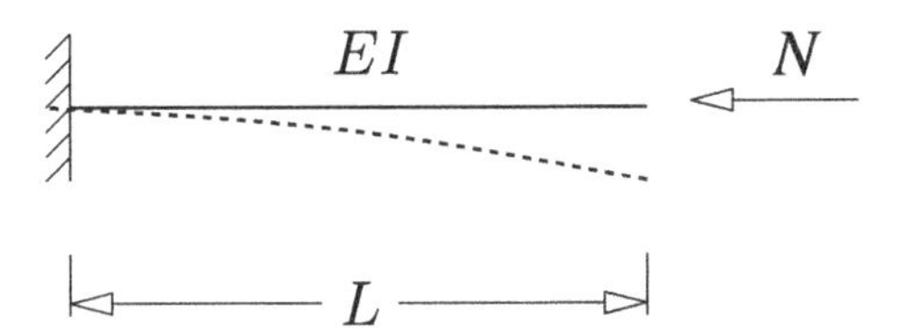

Figure 1.2 Cantilever under axial load.

1.2.3 *Buckling analysis*

Now consider the same cantilever beam with an axial load N located at the end of the beam.

This is a stability problem governed by the differential equation:

$$\frac{d^4w}{dx^4} + \frac{N}{EI}\frac{d^2w}{dx^2} = 0 \tag{1.7}$$

Introducing the parameter λ in terms of

$$\lambda^2 = \frac{N}{EI} \tag{1.8}$$

we can solve this equation by

$$w(x) = A\cos\lambda x + B\sin\lambda x + Cx + D \tag{1.9}$$

Here, the coefficients A, B, C, D have yet to be determined At least one of them should be non-zero in order to obtain a non-trivial solution. From the support conditions on the left end $x = 0$ we easily get:

$$\begin{aligned} w(0) &= 0 \rightarrow A + D = 0 \\ w'(0) &= 0 \rightarrow \lambda B + C = 0 \end{aligned} \tag{1.10}$$

The dynamic boundary conditions are given in terms of the bending moment M at both ends (remember that we need to formulate the equilibrium conditions in the deformed state in order to obtain meaningful results):

$$\begin{aligned} &M(L) = 0 \rightarrow w''(L) = 0 \rightarrow -A\lambda^2\cos\lambda L - B\lambda^2\sin\lambda L = 0 \\ &M(0) = -N\cdot w(L) \rightarrow w''(0) - \frac{N}{EI}w(L) = \\ &\qquad -A\lambda^2 - \lambda^2(A\cos\lambda L + B\sin\lambda L + CL + D) = 0 \\ &\qquad \rightarrow -A\lambda^2(1+\cos\lambda L) - B\lambda^2\sin\lambda L - \lambda^2 CL - \lambda^2 D = 0 \end{aligned} \tag{1.11}$$

Satisfying these four conditions, with at least one of the coefficients being different from zero, requires a singular coefficient matrix, i.e.

$$\det \begin{bmatrix} 1 & 0 & 0 & 1 \\ 0 & \lambda L & 1 & 0 \\ -\lambda^2 \cos\lambda L & -\lambda^2 \sin\lambda L & 0 & 0 \\ \lambda^2 (-\cos\lambda L - 1) & -\lambda^2 \sin\lambda L & -\lambda^2 L & -\lambda^2 \end{bmatrix}$$

$$= \lambda^5 L^2 \cos\lambda L = 0 \rightarrow \lambda L = \frac{(2k-1)\pi}{2}; \; k = 1, \ldots \infty \tag{1.12}$$

Hence the smallest critical load N_{cr}, for which a non-zero equilibrium configuration is possible, is given in terms of λ_1 as

$$N_{cr} = \lambda_1^2 EI = \frac{\pi^2 EI}{4L^2} \tag{1.13}$$

The magnitude of the corresponding deflection remains undetermined. Now assume that $L = 1$ and the load N is a Gaussian random variable with a mean value of 2 and standard deviation of 0.2, and the bending stiffness EI is a Gaussian random variable with a mean value of 1 and standard deviation of 0.1. What is the probability that the actual load N is larger than the critical load N_{cr}?

This `octave` script solves the problem using Monte Carlo simulation.

```
1  M=1000000;
2  N3=2+.2*randn(M,1);
3  EI=1+.1*randn(M,1);
4  Ncr=pi^2*EI/4.;
5  indicator = N>Ncr;
6  pf=mean(indicator)
```

Running this script three times, we obtain (note that again the results do slightly differ in each run)

```
1  pf = 0.070543
2  pf = 0.070638
3  pf = 0.070834
```

In these results, `pf` denotes the mean value of the estimated probability. This problem has an exact solution which can be computed analytically: `pf` $= 0.0705673$. The methods required to arrive at this analytical solution are discussed in chapter 6.

Exercise 1.2 (Buckling)

Consider the same stability problem as above, but now assume that the random variables involved are N, L and EI_0. Presume that these variables have a mean value of 1 and a standard deviation of 0.05. Compute the mean value and the standard

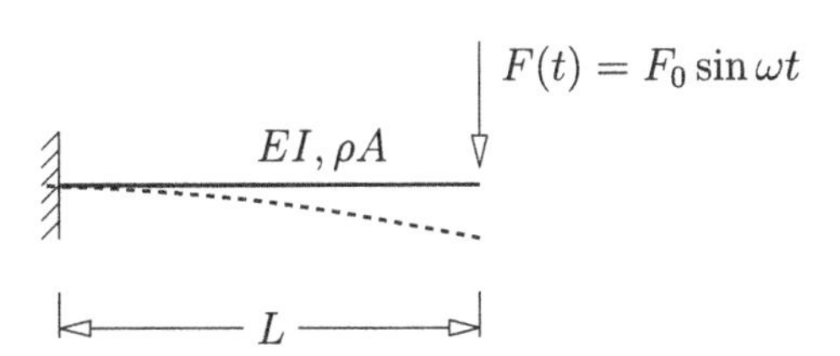

Figure 1.3 Cantilever under dynamic load.

deviation of the critical load applying Monte Carlo simulation using one million samples. Compute the probability that the critical load is less than 2.

Solution: Monte Carlo simulation results in `mn` = 2.4675, `sn` = 0.12324 and `pf` = 9.3000e-05. The last result is not very stable, i.e. it varies quite considerably in different runs. Reasons for this are discussed in chapter 6.

1.2.4 *Dynamic analysis*

Now, we consider the same simple cantilever under a dynamic loading $F(t)$.

For this beam with constant density ρ, cross sectional area A and bending stiffness EI under distributed transverse loading $p(x,t)$, the dynamic equation of motion is

$$\rho A \frac{\partial^2 w}{\partial t^2} + EI \frac{\partial^4 w}{\partial x^4} = p(x,t) \tag{1.14}$$

together with a set of appropriate initial and boundary conditions.

In the following, we would like to compute the probability that the load as given is close to a resonance situation, i.e. the ratio of the excitation frequency ω and the first natural frequency ω_1 of the system is close to 1. The fundamental frequency of the system can be computed from the homogeneous equation of motion:

$$\rho A \frac{\partial^2 w}{\partial t^2} + EI \frac{\partial^4 w}{\partial x^4} = 0 \tag{1.15}$$

by applying the method of separation of variables:

$$w(x,t) = \phi(x) \cdot T(t) \tag{1.16}$$

Inserting this into the differential equation results in

$$\frac{\ddot{T}(t)}{T(t)} = -\frac{EI}{\rho A} \frac{\phi^{IV}(x)}{\phi(x)} = -\omega^2 \tag{1.17}$$

Here the first term is a function only of t, the second term is a function only of x. Obviously, this is only possible if these terms are constants ($= -\omega^2$). Using the right hand side part of Eq. (1.17) we obtain

$$\phi^{IV}(x) - \frac{\rho A}{EI} \omega^2 \phi(x) = 0 \tag{1.18}$$

Introducing the symbol λ defined as

$$\lambda^4 = \frac{\rho A}{EI}\omega^2 \tag{1.19}$$

we can write the general solution of Eq. (1.18) as:

$$\phi(x) = B_1 \sinh \lambda x + B_2 \cosh \lambda x + B_3 \sin \lambda x + B_4 \cos \lambda x \tag{1.20}$$

Here, the constants B_i have to be determined in such a way as to satisfy the kinematic and dynamic boundary conditions at $x = 0$ and $x = L$.

For the cantilever under investigation (fixed end at $x = 0$, free end at $x = L$) the boundary conditions are

$$\phi(0) = 0; \ \phi'(0) = 0; \ \phi''(L) = 0; \ \phi'''(L) = 0 \tag{1.21}$$

Introducing this into (1.20) yields the equations

$$B_2 + B_4 = 0; \ \lambda B_1 + \lambda B_3 = 0; \tag{1.22}$$

as well as

$$\begin{aligned} &\lambda^2 B_1(\sinh \lambda L + \sin \lambda L) + \lambda^2 B_2(\cosh \lambda L + \cos \lambda L) = 0 \\ &\lambda^3 B_1(\cosh \lambda L + \cos \lambda L) + \lambda^3 B_2(\sinh \lambda L - \sin \lambda L) = 0 \end{aligned} \tag{1.23}$$

Non-trivial solutions exist for

$$\cosh \lambda L \cos \lambda L + 1 = 0 \tag{1.24}$$

which has infinitely many positive solutions $\lambda_k; k = 1 \ldots \infty$. The smallest positive solution is $\lambda_1 = \frac{1.875104}{L}$.

Returning to Eq. 1.17 and considering the first part, we obtain

$$\ddot{T}_k + \omega_k^2 T_k = 0 \tag{1.25}$$

This is the well-known differential equation of a single-degree-of-freedom oscillator with the solution

$$T_k(t) = C_{k,1} \cos \omega_k t + C_{k,2} \sin \omega_k t \tag{1.26}$$

In this equation, the constants $C_{k,1}$ and $C_{k,2}$ have to be determined from the initial conditions. The fundamental natural circular frequency ω_1 is therefore given by

$$\omega^2 = \frac{\lambda_1^4 EI}{\rho A} = \frac{12.362 EI}{\rho A L^4} \tag{1.27}$$

Now we assume that the excitation frequency ω is a random variable with a mean value of 0.3 and a standard deviation of 0.03. The bending stiffness is a random variable with mean value 0.1 and standard deviation 0.01, the cross sectional area is random with a mean value of 1 and a standard deviation of 0.05. The density is deterministic $\rho = 1$, so is the length $L = 1$. We want to compute the probability that the ratio $\frac{\omega}{\omega_1}$ lies between 0.99 and 1.01. This is achieved by the following octave script:

```
1  M=1000000;
2  om=1+0.1*randn(M,1);
3  EI=0.1+0.01*randn(M,1);
4  A=1+0.05*randn(M,1);
5  om1=sqrt(EI./A*12.362);
6  ind1 = om./om1>0.99;
7  ind2 = om./om1<1.01;
8  indicator = ind1.*ind2;
9  pr=mean(indicator)
```

Running this script three times, we get

```
1  pr =  0.046719
2  pr =  0.046946
3  pr =  0.046766
```

In these results, pr denotes the mean value of the estimated probability.

Exercise 1.3 (Dynamic deflection)
Now assume that the random variables involved in the above example are A, L and EI. Let these variables have a mean value of 1 and a standard deviation of 0.05. Compute the mean value and the standard deviation of the fundamental natural circular frequency ω_1 using Monte Carlo simulation with one million samples. Compute the probability that ω_1 is between 2 and 2.5.

Solution: Monte Carlo simulation results in the mean value mo = 3.55, the standard deviation so = 0.38 and the probability is of the order of pf = 2.8e-4.

1.2.5 *Structural analysis*

A four-story stucture as sketched in Fig. 1.4 is subjected to four static loads $F_i, i = 1, 2, 3, 4$. The floor slabs are assumed to be rigid and the columns have identical length $H = 4$ m and different bending stiffnesses $EI_k, k = 1 \ldots 8$. Loads and stiffnesses are random variables. The loads are normally distributed with a mean value of 20 kN and a COV of 0.4, the stiffnesses are normally distributed with a mean value of 10 MNm^2 and a COV of 0.2. All variables are pairwise independent.

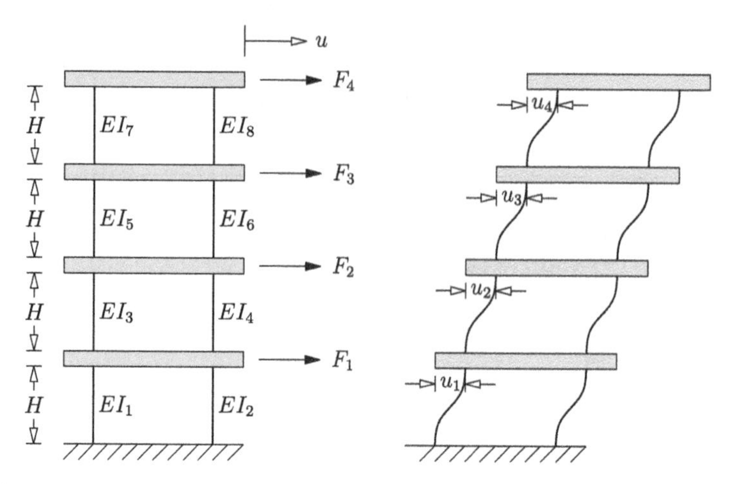

Figure 1.4 Four-story structure under static loads.

We want to compute

- the mean value and standard deviation as well as the coefficient of variation of the horizontal displacement u of the top story,
- the probability p_F that u exceeds a value of 0.1 m.

The analysis is to be based on linear elastic behavior of the structure excluding effects of gravity.

The top story deflection can be calculated by adding the interstory relative displacements (cf. Fig. 1.4):

$$\begin{aligned}
u_4 &= \frac{F_4 H^3}{12(EI_7 + EI_8)} \\
u_3 &= \frac{(F_3 + F_4)H^3}{12(EI_5 + EI_6)} \\
u_2 &= \frac{(F_2 + F_3 + F_4)H^3}{12(EI_3 + EI_4)} \\
u_1 &= \frac{(F_1 + F_2 + F_3 + F_4)H^3}{12(EI_1 + EI_2)} \\
u &= u_1 + u_2 + u_3 + u_4
\end{aligned} \tag{1.28}$$

An `octave`-script carrying out the analysis is shown in Listing 1.1.

```
1  Fbar=20;
2  sigmaF = Fbar*0.4;
3  EIbar=10000;
4  sigmaEI = EIbar*0.2;
5
6  NSIM=1000000;
7  ULIM=0.1;
8  UU=zeros(NSIM,1);
9
10 F1=Fbar + sigmaF*randn(NSIM,1);
11 F2=Fbar + sigmaF*randn(NSIM,1);
12 F3=Fbar + sigmaF*randn(NSIM,1);
13 F4=Fbar + sigmaF*randn(NSIM,1);
14
15 EI1 = EIbar + sigmaEI*randn(NSIM,1);
16 EI2 = EIbar + sigmaEI*randn(NSIM,1);
17 EI3 = EIbar + sigmaEI*randn(NSIM,1);
18 EI4 = EIbar + sigmaEI*randn(NSIM,1);
19 EI5 = EIbar + sigmaEI*randn(NSIM,1);
20 EI6 = EIbar + sigmaEI*randn(NSIM,1);
21 EI7 = EIbar + sigmaEI*randn(NSIM,1);
22 EI8 = EIbar + sigmaEI*randn(NSIM,1);
23
24 H=4;
25
26 u4=F4./(EI7+EI8)/12*H^3;
27 u3=(F3+F4)./(EI5+EI6)/12*H^3;
28 u2=(F2+F3+F4)./(EI3+EI4)/12*H^3;
29 u1=(F1+F2+F3+F4)./(EI1+EI2)/12*H^3;
30 u=u1+u2+u3+u4;
31
32 UM=mean(u)
33 US=std(u)
34 COV=US/UM
35 indic=u>ULIM;
36 PF=mean(indic)
```

Listing 1.1 Monte Carlo simulation of structural analysis.

The results are:

```
1  UM  = 0.054483
2  US  = 0.012792
3  COV = 0.23478
4  PF  = 7.1500e-04
```

Chapter 2

Preliminaries in probability theory and statistics

ABSTRACT: This chapter introduces elementary concepts of probability theory such as conditional probabilities and Bayes' theorem. Random variables and random vectors are discussed together with mathematical models. Some emphasis is given to the modelling of the joint probability density of correlated non-Gaussian random variables using the so-called *Nataf*-model. Statistical concepts and methods are introduced as they are required for sample-based computational methods.

2.1 Definitions

Probability is a measure for the frequency of occurrence of an event. Intuitively, in an experiment this can be explained as the ratio of the number of favorable events to the number of possible outcomes. However, a somewhat more stringent definiton is helpful for a rigorous mathematical foundation (Kolmogorov, see e.g. Papoulis 1984). Axiomatically, this is described by events related to sets $\mathcal{A}, \mathcal{B}, \ldots$ contained in the set Ω, which is the set of all possible events, and a non-negative measure **Prob**(i.e. Probability) defined on these sets following three axioms:

$$\begin{aligned} I: &\quad 0 \leq \mathbf{Prob}[\mathcal{A}] \leq 1 \\ II: &\quad \mathbf{Prob}[\Omega] = 1 \\ III: &\quad \mathbf{Prob}[\mathcal{A} \cup \mathcal{B}] = \mathbf{Prob}[\mathcal{A}] + \mathbf{Prob}[\mathcal{B}] \end{aligned} \tag{2.1}$$

Axiom III holds if $\mathcal{A}$ and $\mathcal{B}$ are mutually exclusive, i.e. $\mathcal{A} \cap \mathcal{B} = \emptyset$.

N.B: The probability associated with the union of two non mutually exclusive events (cf. the example of $\mathcal{A}$ and $\mathcal{C}$ shown in Fig. 2.1) is not equal to the sum of the individual probabilities, $\mathbf{Prob}[\mathcal{A} \cup \mathcal{C}] \neq \mathbf{Prob}[\mathcal{A}] + \mathbf{Prob}[\mathcal{C}]$. In this example there is an apparent overlap of the two events defined by $\mathcal{A} \cap \mathcal{C}$. By removing this overlap, we again obtain mutually exclusive events. From this argument we obtain:

$$\mathbf{Prob}[\mathcal{A} \cup \mathcal{C}] = \mathbf{Prob}[\mathcal{A}] + \mathbf{Prob}[\mathcal{C}] - \mathbf{Prob}[\mathcal{A} \cap \mathcal{C}] \tag{2.2}$$

Given an event $\mathcal{A}$ within the set Ω of all possible events we can define the complementary Event $\bar{\mathcal{A}} = \Omega \backslash \mathcal{A}$ (see Fig. 2.2). Obviously, $\mathcal{A}$ and $\bar{\mathcal{A}}$ are mutually exclusive, hence:

$$\mathbf{Prob}[\mathcal{A}] + \mathbf{Prob}[\bar{\mathcal{A}}] = \mathbf{Prob}[\Omega] = 1 \tag{2.3}$$

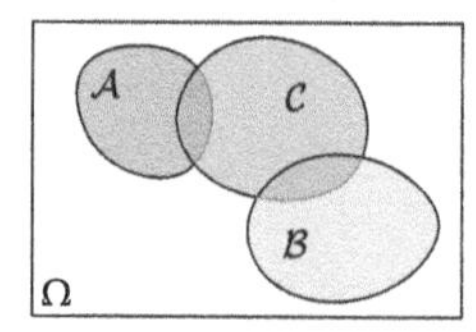

Figure 2.1 Set representation of events in sample space.

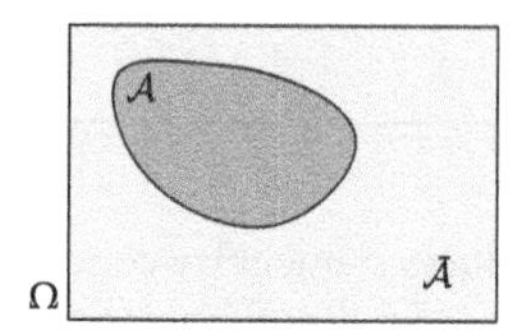

Figure 2.2 Event $\mathcal{A}$ and complementary event $\bar{\mathcal{A}}$.

because of $\textbf{Prob}[A \cap \bar{\mathcal{A}}] = \textbf{Prob}[\emptyset] = 0$. It can be noted that an impossible event has zero probability but the reverse is not necessarily true.

$$\rightarrow \textbf{Prob}[\bar{\mathcal{A}}] = 1 - \textbf{Prob}[\mathcal{A}] \tag{2.4}$$

The conditional probability of an event $\mathcal{A}$ conditional on the occurrence of event $\mathcal{B}$ describes the occurrence probability of $\mathcal{A}$ once we know that $\mathcal{B}$ has already occurred. It can be defined as:

$$\textbf{Prob}[\mathcal{A}|\mathcal{B}] = \frac{\textbf{Prob}[\mathcal{A} \cap \mathcal{B}]}{\textbf{Prob}[\mathcal{B}]} \tag{2.5}$$

Two events $\mathcal{A}$ and $\mathcal{B}$ are called *stochastically independent* if the conditional probability is not affected by the conditioning event, i.e. $\textbf{Prob}[\mathcal{A}|\mathcal{B}] = \textbf{Prob}[\mathcal{A}]$. In this case we have

$$\textbf{Prob}[\mathcal{A} \cap \mathcal{B}] = \textbf{Prob}[\mathcal{A}] \cdot \textbf{Prob}[\mathcal{B}] \tag{2.6}$$

If Ω is partitioned into disjoint sets $A_1 \ldots A_n$ and $\mathcal{B}$ is an arbitrary event (cf. Fig. 2.3), then

$$\textbf{Prob}[\mathcal{B}] = \textbf{Prob}[\mathcal{B}|\mathcal{A}_1] \cdot \textbf{Prob}[\mathcal{A}_1] + \ldots + \textbf{Prob}[\mathcal{B}|\mathcal{A}_n] \cdot \textbf{Prob}[\mathcal{A}_n] \tag{2.7}$$

This is known as the *total probability theorem*. Based on Eq.(2.7) we obtain the so-called *Bayes' theorem*

$$\textbf{Prob}[\mathcal{A}_i|\mathcal{B}] = \frac{\textbf{Prob}[\mathcal{B}|\mathcal{A}_i] \cdot \textbf{Prob}[\mathcal{A}_i]}{\textbf{Prob}[\mathcal{B}|\mathcal{A}_1] \cdot \textbf{Prob}[\mathcal{A}_1] + \ldots + \textbf{Prob}[\mathcal{B}|\mathcal{A}_n] \cdot \textbf{Prob}[\mathcal{A}_n]} \tag{2.8}$$

In this context, the terms *a priori* and *a posteriori* are often used for the probabilities $\textbf{Prob}[\mathcal{A}_i]$ and $\textbf{Prob}[\mathcal{A}_i|\mathcal{B}]$ respectively.

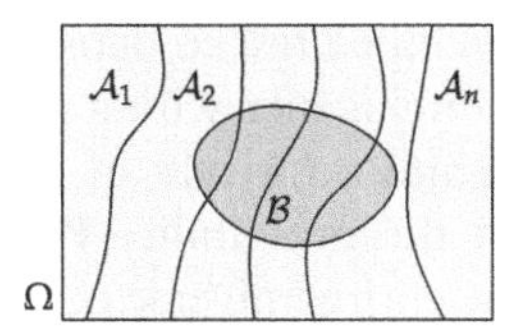

Figure 2.3 Event $\mathcal{B}$ in disjoint partitioning $\mathcal{A}_i$ of Ω.

Example 2.1 (Conditional probabilities)
Consider a non-destructive structural testing procedure to indicate severe structural damage which would lead to imminent structural failure. Assume that the test has the probability $P_{td} = 0.9$ of true detection (i.e. of indicating damage when damage is actually present). Also assume that the test has a probability of $P_{fd} = 0.05$ of false detection (i.e. of indicating damage when damage is actually *not* present). Further assume that the unconditional structural damage probability is $P_D = 0.01$ (i.e. without any test). What is the probability of structural damage if the test indicates positive for damage?

The problem is solved by computing the conditional probability of structural damage given the test is positive. Let structural damage be denoted by the event $\mathcal{A}$ and the positive test result by the event $\mathcal{B}$. A positive test result will occur if

a) the test correctly indicates damage (damage is present)
b) the test falsely indicates damage (damage is not present)

The probabilities associated with these mutually exclusive cases are readily computed as

$$P_a = \textbf{Prob}[\mathcal{A} \cap \mathcal{B}] = \textbf{Prob}[\mathcal{B}|\mathcal{A}] \cdot \textbf{Prob}[\mathcal{A}] = P_{td} \cdot P_D = 0.009$$
$$P_b = \textbf{Prob}[\bar{\mathcal{A}} \cap \mathcal{B}] = \textbf{Prob}[\mathcal{B}|\bar{\mathcal{A}}] \cdot \textbf{Prob}[\bar{\mathcal{A}}] = P_{fd} \cdot (1 - P_D) = 0.0495$$

Hence the probability of a positive test result is

$$\begin{aligned}\textbf{Prob}[\mathcal{B}] &= \textbf{Prob}[\mathcal{A} \cap \mathcal{B}] + \textbf{Prob}[\bar{\mathcal{A}} \cap \mathcal{B}] = \\ &= P_{td} \cdot P_D + P_{fn} \cdot (1 - P_D) = 0.009 + 0.0495 = 0.0585 \end{aligned} \tag{2.9}$$

From this we easily obtain the desired probability according to Bayes' theorem:

$$\begin{aligned}\textbf{Prob}[\mathcal{A}|\mathcal{B}] &= \textbf{Prob}[\mathcal{A} \cap \mathcal{B}]/\textbf{Prob}[\mathcal{B}] = \frac{P_{td} \cdot P_D}{P_{td} \cdot P_D + P_{fn} \cdot (1 - P_D)} \\ &= 0.009/0.0585 = 0.154 \end{aligned} \tag{2.10}$$

This indicates that the test does not perform too well. It is interesting to note that this performance deteriorates significantly with decreasing damage probability P_D, which can easily be seen from the above equation.

Exercise 2.1 (Conditional probability)
Assume that there is a newly developed test to detect extraterrestrial intelligence. This test indicates positively for true intelligence with a probability of $P_{td} = 0.9999$. Unfortunately, the test may also indicate positively in the absence of intelligence with a probability of P_{fd}. Assume that the probability $\mathbf{Prob}[I] = P_I$ of a randomly picked planet carrying intelligence in the entire universe is 0.00001. Compute the limit on the false detection probability P_{fd} so as the conditional probability of intelligence on the condition that the test has a positive result, $\mathbf{Prob}[I|D]$, is larger than 0.5.

Solution: This is to be solved using the relations from the previous example. We get the very small value of $P_{fd} < 9.9991 \cdot 10^{-6} \approx 10^{-5}$.

2.2 Probabilistic models

2.2.1 *Random variables*

For most physical phenomena, random events $\mathcal{A}$ can be suitably defined by the occurrence of a real-valued random value X, which is smaller than a prescribed, deterministic value x.

$$\mathcal{A} = \{X|X < x\} \tag{2.11}$$

The probability $\mathbf{Prob}[\mathcal{A}]$ associated with this event obviously depends on the magnitude of the prescribed value x, i.e. $\mathbf{Prob}[A] = F(x)$ For real valued X and x, this function $F_X(x)$ is called *probability distribution function* (or equivalently, *cumulative distribution function*, cdf)

$$F_X(x) = \mathbf{Prob}[X < x] \tag{2.12}$$

In this notation, the index X refers to the random variable X and the argument x refers to a deterministic value against which the random variable is compared. Since real-valued variables must always be larger than $-\infty$ and can never reach or exceed $+\infty$, we obviously have

$$\lim_{x \to -\infty} F_X(x) = 0; \quad \lim_{x \to +\infty} F_X(x) = 1 \tag{2.13}$$

Differentiation of $F_X(x)$ with respect to x yields the so-called *probability density function* (pdf)

$$f_X(x) = \frac{d}{dx} F_X(x) \tag{2.14}$$

From the above relations it follows that the area under the pdf-curve must be equal to unity:

$$\int_{-\infty}^{\infty} f_X(x) \mathrm{d}x = 1 \tag{2.15}$$

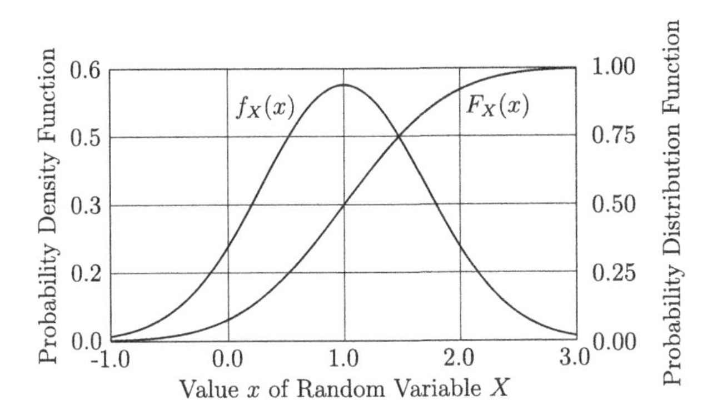

Figure 2.4 Schematic sketch of probability distribution and probability density functions.

A qualitative representation of these relations is given in Fig. 2.4.

In many cases it is more convenient to characterize random variables in terms of expected values rather than probability density functions. The expected value (or ensemble average) of a random quantity $Y = g(X)$ can be defined in terms of the probability density function of X as

$$\mathbf{E}[Y] = \mathbf{E}[g(X)] = \int_{-\infty}^{\infty} g(x) f_X(x) \mathrm{d}x \tag{2.16}$$

From this definition, it is obvious that the expectation operator is linear, i.e.

$$\mathbf{E}[\lambda Y] = \lambda \mathbf{E}[X]; \quad \mathbf{E}[Y + Z] = \mathbf{E}[Y] + \mathbf{E}[Z] \tag{2.17}$$

Special cases of expected values are the *mean value* $\bar{X}$

$$\bar{X} = \mathbf{E}[X] = \int_{-\infty}^{\infty} x f_X(x) \mathrm{d}x \tag{2.18}$$

and the *variance* σ_X^2 of a random variable

$$\sigma_X^2 = \mathbf{E}[(X - \bar{X})^2] = \int_{-\infty}^{\infty} (x - \bar{X})^2 f_X(x) \mathrm{d}x \tag{2.19}$$

The positive square root of the variance σ_X is called *standard deviation*. For variables with non-zero mean value ($\bar{X} \neq 0$) it is useful to define the dimensionless coefficient of variation

$$V_X = \frac{\sigma_X}{\bar{X}} \tag{2.20}$$

A description of random variables in terms of mean value and standard deviation is sometimes called "second moment representation". Note that the mathematical expectations as defined here are so-called *ensemble averages*, i.e. averages over all possible realizations.

A generalization of these relations is given by the definition of statistical moments of k-th order

$$\mu_k = \mathbf{E}[X^k] = \int_{-\infty}^{\infty} x^k f_X(x)\mathrm{d}x \tag{2.21}$$

and the centralized statistical moments of k-th order

$$\hat{\mu}_k = \mathbf{E}[(X - \bar{X})^k] = \int_{-\infty}^{\infty} (x - \bar{X})^k f_X(x)\mathrm{d}x \tag{2.22}$$

In some applications, two specific normalized statistical moments are of interest. These dimensionless quantities are the skewness s defined by

$$s = \frac{\hat{\mu}_3}{\sigma_X^3} \tag{2.23}$$

and the kurtosis (or excess) κ defined by

$$\kappa = \frac{\hat{\mu}_4}{\sigma_X^4} - 3 \tag{2.24}$$

Note that for a Gaussian distribution both skewness and kurtosis are zero.

Theorem: (**Chebyshev's inequality**)
Assume X to be a random variable with a mean value of $\bar{X}$ and finite variance $\sigma_X^2 < \infty$. Then:

$$\mathbf{Prob}[|X - \bar{X}| > \epsilon] \leq \frac{\sigma_X^2}{\epsilon^2} \qquad \forall \epsilon > 0 \tag{2.25}$$

For many practical applications (such as, e.g. in structural reliability analysis) this bound is not sufficiently narrow. For example, consider the case in which $\epsilon = \sigma_X$. From Chebyshev's inequality we obtain the result $\mathbf{Prob}[|X - \bar{X}| \geq \sigma_X] \leq 1$. This result is not really helpful.

Standardization

This is a linear transformation of the original variable X to a new variable Y which has zero mean and unit standard deviation.

$$Y = \frac{X - \bar{X}}{\sigma_X} \tag{2.26}$$

Based on the linearity of the expectation operator (cf. Eq. 2.17) it is readily shown that the mean value of Y is zero

$$\mathbf{E}[Y] = \frac{1}{\sigma_X}(\mathbf{E}[X] - \mathbf{E}[\bar{X}]) = 0 \tag{2.27}$$

and that the standard deviation is equal to unity

$$\mathbf{E}[Y^2] = \frac{1}{\sigma_X^2}\mathbf{E}[(X - \bar{X})^2] = \frac{\sigma_X^2}{\sigma_X^2} = 1 \tag{2.28}$$

Y is called a *standardized* random variable.

2.2.2 *Some types of distributions*

Gaussian distribution

Due to its simplicity, the so-called Gaussian or normal distribution is frequently used. A random variable X is *normally distributed* if its probability density function is:

$$f_X(x) = \frac{1}{\sqrt{2\pi}\sigma_X}\exp\left[-\frac{(x - \bar{X})^2}{2\sigma_X^2}\right]; \quad -\infty < x < \infty \tag{2.29}$$

Here, $\bar{X}$ is the mean value and σ_X is the standard deviation. The distribution function $F_X(x)$ is described by the normal integral $\Phi(.)$:

$$F_X(x) = \Phi\left(\frac{x - \bar{X}}{\sigma_X}\right) \tag{2.30}$$

in which

$$\Phi(z) = \frac{1}{\sqrt{2\pi}}\int_{-\infty}^{z}\exp\left(-\frac{u^2}{2}\right)\mathrm{d}u \tag{2.31}$$

This integral is not solvable in closed form, however, tables and convenient numerical approximations exist. The use of the Gaussian distribution is frequently motivated by the central limit theorem, which states that an additive superposition of independent random effects of similar importance tends asymptotically to the Gaussian distribution.

For the Gaussian distribution, we can easily derive expressions for the higher-order statistical moments in terms of the mean value and the standard deviation. For the third-order moment we get

$$\mathbf{E}[X^3] = \int_{-\infty}^{\infty} x^3 \frac{1}{\sqrt{2\pi}\sigma_X} \exp\left[-\frac{(x-\bar{X})^2}{2\sigma_X^2}\right] dx$$

$$= \int_{-\infty}^{\infty} (u\sigma_x + \bar{X})^3 \frac{1}{\sqrt{2\pi}} \exp\left(-\frac{u^2}{2}\right) du = 3\sigma_X^3 + \bar{X}^3 \tag{2.32}$$

and for the fourth-order moment

$$\mathbf{E}[X^4] = \int_{-\infty}^{\infty} (u\sigma_x + \bar{X})^4 \frac{1}{\sqrt{2\pi}} \exp\left(-\frac{u^2}{2}\right) du$$

$$= 3\sigma_X^4 + 6\sigma_X^2\bar{X}^2 + \bar{X}^4 \tag{2.33}$$

Log-normal distribution

A random variable X is *log-normally* distributed if its probability density function is

$$f_X(x) = \frac{1}{x\sqrt{2\pi}s} \exp\left[-\frac{(\log\frac{x}{\mu})^2}{2s^2}\right]; \; x \geq 0 \tag{2.34}$$

and its distribution function is given by

$$F_X(x) = \Phi\left(\frac{\log\frac{x}{\mu}}{s}\right) \tag{2.35}$$

In these equations, the parameters μ and s are related to the mean value and the standard deviation as follows:

$$s = \sqrt{\ln\left(\frac{\sigma_X^2}{\bar{X}^2} + 1\right)}; \quad \mu = \bar{X}\exp\left(-\frac{s^2}{2}\right) = \frac{\bar{X}^2}{\sqrt{\bar{X}^2 + \sigma_X^2}} \tag{2.36}$$

Two random variables with $\bar{X} = 1.0$ and $\sigma_X = 0.5$ that have different distribution types are shown in Fig. 2.5. It can clearly be seen that the log-normal density function is non-symmetric.

The difference becomes significant especially in the tail regions. The log-normal distribution does not allow any negative values, whereas the Gaussian distribution, in this case, allows negative values with a probability of 2.2%. In the upper tail, the probabilities of exceeding different threshold values ξ, i.e. $\mathbf{Prob}[X > \xi]$, are shown in Table 2.1. The difference is quite dramatic what underlines the importance of the appropriate distribution model.

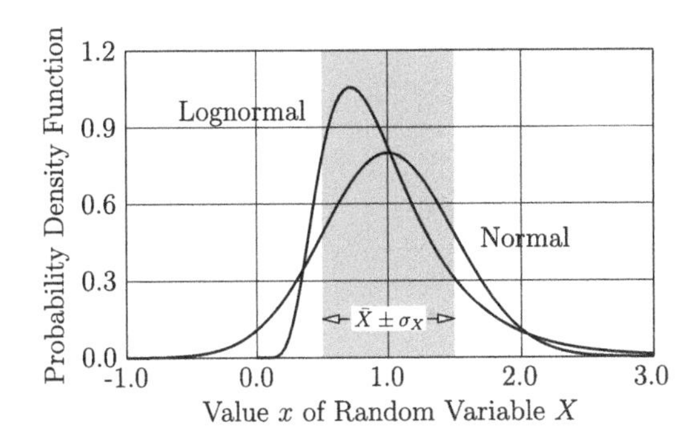

Figure 2.5 Normal and log-normal probability density functions.

Table 2.1 Exceedance probabilities for different distribution models.

Level ξ	*0*	*1*	*2*	*3*	*4*	*5*
Normal	0.97725	0.5	0.0228	3.17e-05	9.90e-10	6.28e-16
Lognormal	1.0	0.40664	0.0442	0.052	0.00076	0.00013

Exponential and Gamma distributions

The family of Gamma distributions is characterized by the probability density function

$$f_X(x) = \frac{1}{a^\nu \Gamma(\nu)} x^{\nu-1} \exp\left(-\frac{x}{a}\right); \quad x \geq 0 \tag{2.37}$$

Here ν is a shape parameter, a is a scale parameter, and $\Gamma(.)$ is the complete Gamma function. The mean value is $\bar{X} = \nu a$ and the standard deviation is given by $\sigma_X = \sqrt{\nu} a$.

For the special case $\nu = 1$ we obtain the exponential density function

$$f_X(x) = \frac{1}{a} \exp\left(-\frac{x}{a}\right); \quad x \geq 0 \tag{2.38}$$

This density function can also be shifted by an amount ϵ so that

$$f_X(x) = \frac{1}{a} \exp\left(-\frac{x-\epsilon}{a}\right); \quad x \geq \epsilon \tag{2.39}$$

Gumbel distribution

The Gumbel (or Type I largest) distribution is frequently used to model extreme events such as annual maximum wind speeds or flood heights. Its probability density function is

$$f_X(x) = \alpha \exp\{-(x-\mu)\alpha - \exp[-(x-\mu)\alpha]\} \tag{2.40}$$

and its probability distribution function is given by

$$F_X(x) = \exp\{-\exp[-(x-\mu)\alpha]\} \tag{2.41}$$

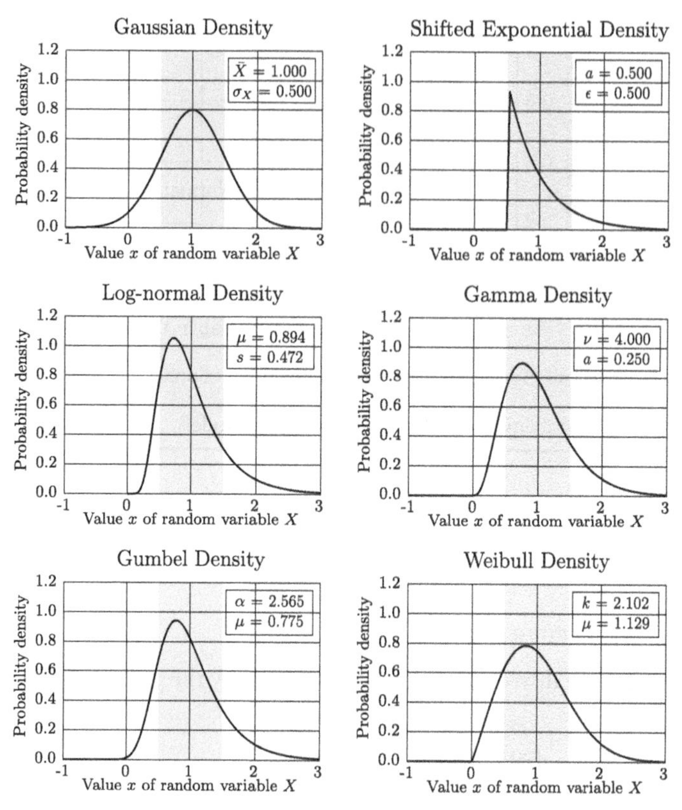

Figure 2.6 Various probability density functions with $\bar{X}=1$ and $\sigma_X=0.5$ (Ranges of mean value $\pm$ standard deviation are indicated as shaded areas).

The mean value and the standard deviation of a Gumbel-distributed random variable X are related to the distribution parameters by

$$\bar{X} = \mu + \frac{\gamma}{\alpha}; \quad \sigma_X = \frac{\pi}{\alpha\sqrt{6}} \tag{2.42}$$

in which $\gamma = 0.5772156649\ldots$ is the Euler-Mascheroni constant (Abramowitz and Stegun 1970).

Weibull distribution

The Weibull (or Type III smallest) distribution is often used in the modeling of material defects related to fatigue and similar problems. Its probability density function is

$$f_X(x) = \frac{k}{\mu}\left(\frac{x}{\mu}\right)^{k-1} \exp\left[-\left(\frac{x}{\mu}\right)^k\right]; \quad x \geq 0 \tag{2.43}$$

Here $\mu > 0$ is a scale parameter and $k > 0$ a shape parameter. The probability distribution function is

$$F_X(x) = 1 - \exp\left[-\left(\frac{x}{\mu}\right)^k\right]; \quad x \geq 0 \tag{2.44}$$

The mean value and the standard deviation of a Weibull-distributed random variable X are related to the distribution parameters μ and k by

$$\bar{X} = \mu \cdot \Gamma\left(1 + \frac{1}{k}\right); \quad \sigma_X^2 = \mu^2\left[\Gamma\left(1 + \frac{2}{k}\right) - \Gamma^2\left(1 + \frac{1}{k}\right)\right] \tag{2.45}$$

Here $\Gamma(.)$ denotes the Gamma function. The limiting case $k = 1$ defines an exponential distribution, the case $k = 2$ is called *Rayleigh*-distribution. The computation of k and μ given $\bar{X}$ and σ_X requires a numerical procedure. Eqs. 2.45 can be reformulated such that μ is eliminated:

$$\frac{\sigma_X^2}{\bar{X}^2} = \frac{\Gamma\left(1 + \frac{2}{k}\right) - \Gamma^2\left(1 + \frac{1}{k}\right)}{\Gamma^2\left(1 + \frac{1}{k}\right)} \to 1 + \frac{\sigma_X^2}{\bar{X}^2} = \frac{\Gamma\left(1 + \frac{2}{k}\right)}{\Gamma^2\left(1 + \frac{1}{k}\right)} \tag{2.46}$$

This can easily be solved, e.g. by bisection. Then μ is given by

$$\mu = \frac{\bar{X}}{\Gamma\left(1 + \frac{1}{k}\right)} \tag{2.47}$$

Example 2.2 (Computation of Weibull parameters)
Let X be a Weibull-distributed random variable with mean value $\overline{X} = 1$ and standard deviation $\sigma_X = 0.5$. Compute the parameters k and μ of the Weibull distribution. The equation to be solved is

$$\frac{\Gamma\left(1 + \frac{2}{k}\right)}{\Gamma^2\left(1 + \frac{1}{k}\right)} - 1.25 = 0 \tag{2.48}$$

The `octave` script solving the example by bisection is shown in Listing 2.1.

It produces the result

```
1 k = 2.1013
2 mu = 1.1291
```

There is also a shifted Weibull (or three-parameter) distribution with an additional parameter ϵ defining a lower limit. In this case, the probability distribution function is

$$F_X(x) = 1 - \exp\left[-\left(\frac{x - \epsilon}{\mu}\right)^k\right]; \quad x \geq \epsilon \tag{2.49}$$

The probability density function follows from differentiation.

```
1  xbar = 1;
2  sigma = 0.5;
3  cc = 1+sigma ^2/xbar ^2;
4
5  k=100;
6  dk=100;
7  while (dk>1e-5)
8     h = gamma(1+2/k)/gamma(1+1/k) ^2 - cc;
9     dk = dk/2;
10    if (h<0) k = k - dk;
11    else k = k + dk;
12      end
13      end
14  k
15  mu=xbar/gamma(1+1/k)
```

Listing 2.1 Computation of Weibull distribution parameters using bisection.

Cauchy distribution

An interesting class of probability density functions is given by rational functions (i.e. fractions of polynomials in x). One classical example is the Cauchy distribution with the probability distribution function

$$F_X(x) = \frac{1}{2} + \frac{1}{\pi}\arctan(x) \tag{2.50}$$

and the probability density function

$$f_X(x) = \frac{1}{\pi(1+x^2)} \tag{2.51}$$

All distributions based on a rational density function only have a finite number of moments (cf. Eq. 2.21). Probability density functions with such properties occur in the analysis of dynamic systems with random parametric excitation (see e.g. Arnold and Imkeller 1994). The Cauchy distribution may be assigned a mean value of zero, although, strictly speaking, the integral

$$\begin{aligned}\bar{X} &= \int_{-\infty}^{\infty} \frac{x\,dx}{\pi(1+x^2)} = \lim_{v\to\infty}\int_{-v}^{0} \frac{x\,dx}{\pi(1+x^2)} + \lim_{u\to\infty}\int_{0}^{u} \frac{x\,dx}{\pi(1+x^2)} \\ &= -\lim_{v\to\infty}\frac{1}{2\pi}\log(1+v^2) + \lim_{u\to\infty}\frac{1}{2\pi}\log(1+u^2)\end{aligned} \tag{2.52}$$

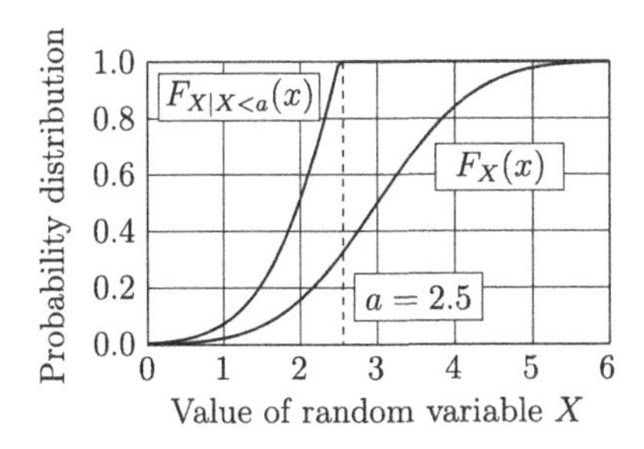

Figure 2.7 Schematic sketch of conditional probability distribution function.

does not exist. Its so-called principal value (which is obtained by letting $u=v$ before taking the limit in the above equation), however, is zero. The variance σ_X^2 is clearly unbounded.

2.2.3 *Conditional distribution*

The concept of conditional probabilities as outlined earlier can readily be extended to the case of random variables. Let $\mathcal{A}$ be the event $X < x$ and $\mathcal{B}$ be the event $X < a$. The probability of $X < x$, given that we already know $X < a$, is then a conditional distribution, i.e.

$$\textbf{Prob}[X < x|X < a] = F_{X|X<a}(x) \tag{2.53}$$

Since this is a probability measure, we obviously have the limits

$$\lim_{x \to -\infty} F_{X|X<a}(x) = 0; \quad \lim_{x \to +\infty} F_{X|X<a}(x) = 1 \tag{2.54}$$

If we assume that the number a is chosen such that $F_X(a) \neq 0$ we can then use the definition of conditional probabilities to compute

$$F_{X|X<a}(x) = \textbf{Prob}[X < x|X < a] = \frac{\textbf{Prob}[(X < x) \wedge (X < a)]}{\textbf{Prob}[X < a]} \tag{2.55}$$

Here we need to distinguish two cases. First, if $x \geq a$ then the set of all X with $X < a$ is a subset of the set of all X with $X < x$. Hence

$$F_{X|X<a}(x) = \frac{\textbf{Prob}[(X < x) \wedge (X < a)]}{\textbf{Prob}[X < a]} = \frac{\textbf{Prob}[X < a]}{\textbf{Prob}[X < a]} = 1 \tag{2.56}$$

Second, if $x < a$ then the set of all X with $X < x$ is a subset of the set of all X with $X < a$. In this case, we obtain (cf. Fig. 2.7)

$$F_{X|X<a}(x) = \frac{\textbf{Prob}[(X < x) \wedge (X < a)]}{\textbf{Prob}[X < a]} = \frac{\textbf{Prob}[X < x]}{\textbf{Prob}[X < a]} = \frac{F_X(x)}{F_X(a)} \tag{2.57}$$

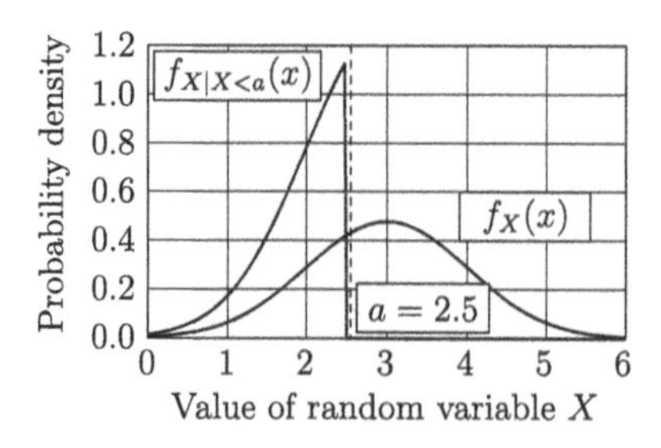

Figure 2.8 Schematic sketch of conditional probability density function.

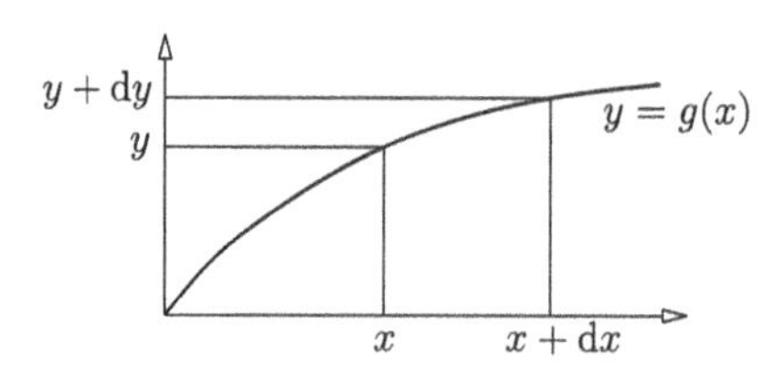

Figure 2.9 Nonlinear function of a random variable.

This can easily be extended to the definition of conditional probability density functions (cf. Fig. 2.8)

$$f_{X|X<a} = \frac{\mathrm{d}}{\mathrm{d}x} F_{X|X<a} = \frac{f_X(x)}{F_X(a)} \tag{2.58}$$

2.2.4 *Functions of random variables*

Let $y = g(x)$ be a strictly monotonic function of x. Then g can be uniquely inverted and the inverse function $x = g^{-1}(y)$ is strictly monotonic too. Then the probability density function of the random variable $Y = g(X)$ is given by

$$f_Y(y) = f_X[g^{-1}(y)] \left| \frac{\mathrm{d}[g^{-1}(y)]}{\mathrm{d}y} \right| \tag{2.59}$$

The proof is based on the consideration that the probability of obtaining a value of Y in the interval $[y = g(x), y + \mathrm{d}y = g(x + \mathrm{d}x)]$ is equal to the probability of obtaining a value of X in the interval $[x, x + \mathrm{d}x]$ (cf. Fig. 2.9) For monotonically increasing $g(.)$ we have

$$\begin{aligned} \mathbf{Prob}[y \leq Y \leq y + \mathrm{d}y] &= \mathbf{Prob}[x \leq X \leq x + \mathrm{d}x] = f_Y(y)\mathrm{d}y = f_X(x)\mathrm{d}x \\ &\rightarrow f_Y(y) = f_X(x)\frac{\mathrm{d}x}{\mathrm{d}y} \end{aligned} \tag{2.60}$$

whereas for monotonically decreasing $g(.)$ we get

$$\textbf{Prob}[y \leq Y \leq y + \mathrm{d}y] = \textbf{Prob}[x + \mathrm{d}x \leq X \leq x] = f_Y(y)\mathrm{d}y = -f_X(x)\mathrm{d}x$$
$$\rightarrow f_Y(y) = -f_X(x)\frac{\mathrm{d}x}{\mathrm{d}y} \tag{2.61}$$

Putting these two equations together and inverting $g(x) = y$ yields the result in Eq. 2.59. If the function $g(.)$ is not monotonic, then the same type of argument can be applied by considering pieces of the function over which it is monotonic.

Example 2.3 (Quadratic function)
Consider the function $g(x) = x^2$. This function is monotonic in the intervals $(-\infty, 0]$ and $[0, \infty)$. We have two possible and mutually exclusive inversions of $g(.)$, namely $x = \sqrt{y}$ and $x = -\sqrt{y}$. The probability density function of Y can then be expressed as

$$f_Y(y) = f_X(\sqrt{y})\frac{1}{2\sqrt{y}} + f_X(-\sqrt{y})\left|-\frac{1}{2\sqrt{y}}\right| = \frac{1}{2\sqrt{y}}[f_X(\sqrt{y}) + f_X(-\sqrt{y})] \tag{2.62}$$

For a Gaussian variable X with mean value of 1 and a standard deviation of 0.1, we get the probability density function of $Y = X^2$ as

$$f_Y(y) = \frac{1}{2\sqrt{y}}\frac{1}{0.1 \cdot \sqrt{2\pi}}\left[\exp\left(-\frac{(\sqrt{y}-1)^2}{0.02}\right) + \exp\left(-\frac{(-\sqrt{y}-1)^2}{0.02}\right)\right] \tag{2.63}$$

This function is shown in Fig. 2.10. This figure shows that the mean value $\bar{Y}$ is close to $\bar{X} = 1$ and that the standard deviation σ_Y is significantly larger than the standard deviation $\sigma_X = 0.1$. Elementary computation shows that

$$\begin{aligned}
\bar{Y} &= \textbf{E}[Y] = \textbf{E}[X^2] = \bar{X}^2 + \sigma_X^2 = 1 + 0.01 = 1.01 \\
\sigma_Y^2 &= \textbf{E}[Y^2] - \textbf{E}[Y]^2 = \textbf{E}[X^4] - (\bar{X}^2 + \sigma_X^2)^2 \\
&= \bar{X}^4 + 6\bar{X}^2\sigma_X^2 + 3\sigma_X^4 - \bar{X}^4 - 2\bar{X}^2\sigma_X^2 - \sigma_X^4 \\
&= 4\bar{X}^2\sigma_X^2 + 2\sigma_X^4 = 0.04 + 0.0002 = 0.0402 \\
\sigma_Y &= 0.2005 \approx 2\sigma_X
\end{aligned} \tag{2.64}$$

For these relations, cf. Eqs. 2.32 and 2.33.

Actually, the probability density function of Y has a singularity at $y = 0$ due to the singularity of the derivative of the inverse transformation $\frac{\mathrm{d}x}{\mathrm{d}y} = \frac{1}{2\sqrt{y}}$ at $y = 0$. However, this singularity (clipped in Fig. 2.10) is insignificant since it does not lead to a finite probability of Y becoming zero.

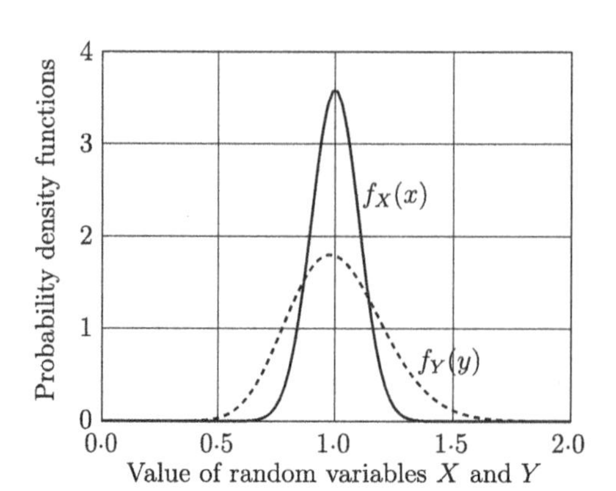

Figure 2.10 PDF of a Gaussian variable X and its square $Y = X^2$.

2.2.5 *Random vectors*

In many applications, a large number of random variables occur together. It is conceptually helpful to assemble all these random variables X_k; $k = 1 \dots n$ into a *random vector* $\mathbf{X}$:

$$\mathbf{X} = [X_1, X_2, \dots X_n]^T \tag{2.65}$$

For this vector, the expected value can be defined in terms of the expected values of all its components:

Mean value vector

$$\bar{\mathbf{X}} = \mathbf{E}[\mathbf{X}] = [\bar{X_1}, \bar{X_2}, \dots \bar{X_n}]^T \tag{2.66}$$

This definition applies the expectation operator (ensemble average) to each component of $\mathbf{X}$ individually.

Covariance matrix

$$\mathbf{E}[(\mathbf{X} - \bar{\mathbf{X}})(\mathbf{X} - \bar{\mathbf{X}})^T] = \mathbf{C}_{\mathbf{XX}} \tag{2.67}$$

This definition means that the expectation operator is applied to all possible mixed products of the zero mean components $(X_i - \bar{X}_i)(X_k - \bar{X}_k)$. As a consequence of this definition, the covariance matrix $\mathbf{C}_{\mathbf{XX}}$ is obviously symmetric. In addition, it is non-negative definite (i.e. it does not have any negative eigenvalues). Therefore, it can be factored by a Cholesky-decomposition

$$\mathbf{C}_{\mathbf{XX}} = \mathbf{L}\mathbf{L}^{\mathrm{T}} \tag{2.68}$$

in which $\mathbf{L}$ is a non-singular lower triangular matrix. The Cholesky factor $\mathbf{L}$ can be utilized for a representation of the random variables X_i in terms of zero-mean uncorrelated random variables Y_i by applying a linear transfomation:

$$\mathbf{Y} = \mathbf{L}^{-1}(\mathbf{X} - \bar{\mathbf{X}}); \quad \mathbf{X} = \mathbf{L}\mathbf{Y} + \bar{\mathbf{X}} \tag{2.69}$$

We can easily prove that the mean value vector of $\mathbf{Y}$ is zero

$$\mathbf{E}[\mathbf{Y}] = \mathbf{E}[\mathbf{L}^{-1}(\mathbf{X} - \bar{\mathbf{X}})] = \mathbf{L}^{-1}\mathbf{E}[\mathbf{X} - \bar{\mathbf{X}}] = 0 \tag{2.70}$$

and that the components of $\mathbf{Y}$ are uncorrelated:

$$\begin{aligned} \mathbf{E}[\mathbf{Y}\mathbf{Y}^T] &= \mathbf{C}_{\mathbf{YY}} = \mathbf{E}[\mathbf{L}^{-1}(\mathbf{X} - \bar{\mathbf{X}})(\mathbf{X} - \bar{\mathbf{X}})^T\mathbf{L}^{-1T}] \\ &= \mathbf{L}^{-1}\mathbf{L}\mathbf{L}^T\mathbf{L}^{-1T} = \mathbf{I} \\ &\rightarrow \mathbf{E}[Y_i^2] = 1 \quad \forall i; \quad \mathbf{E}[Y_i Y_k] = 0 \quad \forall i \neq k \end{aligned} \tag{2.71}$$

The dimensionless quantity

$$\rho_{ik} = \frac{\mathbf{E}[(X_i - \bar{X}_i)(X_k - \bar{X}_k)]}{\sigma_{X_i}\sigma_{X_k}} \tag{2.72}$$

is called *coefficient of correlation*. Its value is bounded in the interval $[-1, 1]$. Note that the matrix of correlation coefficients must be positive definite as well. This poses certain restrictions on the numerical values of ρ_{ik} depending on the dimension of the random vector $\mathbf{X}$.

Example 2.4 (Simulation of correlated random variables)
Assume that the area of a rectangle is given in terms of two correlated Gaussian random variables X_1 and X_2, $A = X_1 X_2$. Both variables have mean values of 1 and standard deviations of 0.5. Assume that they are correlated with $\rho_{12} = 0.7$.

a) Derive a linear transformation to generate samples of X_1 and X_2. This is done by performing a Cholesky decomposition on the covariance matrix.
b) Compute the mean value and the standard deviation of the rectangle area A. This is done by carrying out a Monte Carlo simulation using correlated variables as input and repeated computation of the area A. The correlated variables are generate from zero mean, standardized uncorrelated variables by multiplying with the Cholesky factor of the covariance matrix and finally adding the mean value.

The `octave` script solving the example is shown in Listing 2.2.

```
1  M=1000000;
2  rho=0.7;
3  Cxx=[0.25, rho*0.25;
4  rho*0.25, 0.25];
5  L=chol(Cxx)';
6  y1=randn(M,1);
7  y2=randn(M,1);
8  x=L*[y1,y2];
9  x1=x(1,:) +1;
10 x2=x(2,:) +1;
11 area=x1.*x2;
12 ma=mean(area)
13 sa=std(area)
```

Listing 2.2 Simulation of correlated random variables.

Running this script we obtain

```
1  ma = 1.1755
2  sa = 0.97135
```

In these results, ma denotes the mean value and sa the standard deviation.

Exercise 2.2 (Random vectors)
Consider a random vector $\mathbf{X}$ containing the three correlated Gaussian random variables X_1, X_2, and X_3. All three variables have a mean value of $\bar{X}=1$ and a standard deviation of $\sigma_X=0.2$. The coefficients of correlation are $\rho_{12}=0.5$, $\rho_{13}=0.2$, $\rho_{23}=-0.4$. Derive a transformation to zero-mean uncorrelated random variables Y_k. Utilize this transformation to perform a Monte Carlo simulation of the random vector $\mathbf{X}$. Assume that the components of $\mathbf{X}$ define the side lengths of a cuboid. Compute the mean value and the standard deviation of the volume V of the cuboid. Furthermore, compute the probability that the volume exceeds that value 2.0. Finally, change one coefficient of correlation $\rho_{13}=0.6$. What do you observe? How can you explain this?

Solution: The mean value of the volume is vm $=1.013$, the standard deviation is vs $=0.387$. The probability of having a volume larger than 2 is pf $=0.0163$. For $\rho_{13}=0.6$ the correlation matrix is not positive definite.

2.2.6 *Joint probability density function models*

Multi-dimensional Gaussian distribution

The pdf of jointly normally (Gaussian) distributed random variables (components of a random vector $\mathbf{X}$ is given by

$$f_{\mathbf{X}}(\mathbf{x}) = \frac{1}{(2\pi)^{\frac{n}{2}}\sqrt{\det \mathbf{C}_{\mathbf{XX}}}} \exp\left[-\frac{1}{2}(\mathbf{x}-\bar{\mathbf{X}})^T \mathbf{C}_{\mathbf{XX}}^{-1}(\mathbf{x}-\bar{\mathbf{X}})\right]; \quad \mathbf{x} \in \mathbb{R}^n \tag{2.73}$$

Independent random variables

If all random variables X_i are mutually independent, then the joint probability density function is given by the product of the individual probability density functions.

$$f_{\mathbf{X}}(\mathbf{x}) = \prod_{i=1}^{n} f_{X_i}(x_i) \tag{2.74}$$

This follows from the multiplication rule for independent events.

Remarks:

1. Independent random variables are always uncorrelated. The reverse is not necessarily true.
2. If the random variables $X_i, i = 1 \ldots n$ are jointly normally distributed and they are pairwise uncorrelated, then they are pairwise independent.
   Proof: From $C_{ik} = 0; i \neq k$ we have

$$\rightarrow (\mathbf{x} - \bar{\mathbf{X}})^T \mathbf{C}_{\mathbf{XX}}^{-1} (\mathbf{x} - \bar{\mathbf{X}}) = \sum_{i=1}^{n} \left(\frac{x_i - \bar{X}_i}{\sigma_{X_i}} \right)^2; \quad \det \mathbf{C}_{\mathbf{XX}} = \prod_{i=1}^{n} \sigma_{X_i}^2 \tag{2.75}$$

Due to a fundamental property of the exponential function we have

$$f_{\mathbf{X}}(\mathbf{x}) = \frac{1}{(2\pi)^{\frac{n}{2}}} \frac{1}{\prod_{i=1}^{n} \sigma_{X_i}} \exp\left[-\frac{1}{2} \sum_{i=1}^{n} \left(\frac{x_i - \bar{X}_i}{\sigma_{X_i}} \right)^2 \right]$$
$$= \prod_{i=1}^{n} \frac{1}{\sqrt{2\pi}\sigma_{X_i}} \exp\left[-\frac{1}{2} \left(\frac{x_i - \bar{X}_i}{\sigma_{X_i}} \right)^2 \right] = \prod_{i=1}^{n} f_{X_i}(x_i) \tag{2.76}$$

which concludes the proof.

Nataf-model

The so-called *Nataf*-model (Nataf 1962, Liu and DerKiureghian 1986) describes the joint probability density function of random variables X_i based on their individual (marginal) distributions and the covariances or coefficients of correlation ρ_{ik}. The concept of this Gaussian copula (Noh, Choi, and Du 2008) is to transform the original variables X_i to Gaussian variables Y_i whose joint density is assumed to be multi-dimensional Gaussian. This model can be realized in three steps:

1. Map all random variables X_i individually to normally distributed random variables V_i with zero mean and unit standard deviation

$$\{X_i; f_{X_i}(x_i)\} \leftrightarrow \{V_i; \varphi(v_i)\} \tag{2.77}$$

which is accomplished by means of

$$V_i = \Phi^{-1}[F_{X_i}(X_i)] \tag{2.78}$$

2. Assume a jointly normal distribution for all random variables V_i with the statistical moments

$$\mathbf{E}[V_i] = 0; \quad \mathbf{E}[V_i^2] = 1; \quad \mathbf{E}[V_i V_k] = \rho'_{ik} \tag{2.79}$$

Note that at this point, the correlation coefficient ρ'_{ik} (which generally will be different from ρ_{ik}) is not yet known. The joint pdf for the components of the random vector $\mathbf{V}$ is then

$$f_{\mathbf{V}}(\mathbf{v}) = \frac{1}{(2\pi)^{\frac{n}{2}}\sqrt{\det \mathbf{R}_{\mathbf{VV}}}} \exp\left(-\frac{1}{2}\mathbf{v}^T \mathbf{R}_{\mathbf{VV}}^{-1} \mathbf{v}\right) \tag{2.80}$$

in which $\mathbf{R}_{\mathbf{VV}}$ denotes the matrix of all correlations ρ'_{ik}. From this relation, it follows that

$$f_{\mathbf{X}}(\mathbf{x}) = f_{\mathbf{V}}[\mathbf{v}(\mathbf{x})] \prod_{i=1}^{n} \left|\frac{dx_i}{dv_i}\right| = f_{\mathbf{V}}[\mathbf{v}(\mathbf{x})] \prod_{i=1}^{n} \frac{f_{X_i}(x_i)}{\varphi[v_i(x_i)]} \tag{2.81}$$

3. Compute the correlation coefficients ρ'_{ik} by solving

$$\sigma_{X_i}\sigma_{X_j}\rho_{ik} = \int_{-\infty}^{\infty}\int_{-\infty}^{\infty} (x_i - \bar{X}_i)(x_k - \bar{X}_k) f_{X_i X_k}(x_i, x_k, \rho'_{ik}) \mathrm{d}x_i \, \mathrm{d}x_k \tag{2.82}$$

This is usually achieved by iteration.

A known problem of the Nataf-model is that this iteration may lead to a non-positive-definite matrix of correlation coefficients. In this case, this model is not applicable. A set of semi-empirical formulas relating ρ and ρ' based on numerical studies for various types of random variables is given by Liu and DerKiureghian 1986.

Example 2.5 (Nataf model for two log-normally distributed random variables) Consider two log-normally distributed random variables X_1, X_2 with identical means $\bar{X}_1 = \bar{X}_2 = \bar{X} = 1$ and identical standard deviations $\sigma_1 = \sigma_2 = \sigma = 0.4$. Assume further that the two variables are correlated with a coefficient of correlation ρ. The individual variables are readily mapped to standard Gaussian variables V_1, V_2 by means of

$$V_i = \Phi^{-1}[F_{X_i}(X_i)] = \Phi^{-1}\Phi\left(\frac{\log \frac{X_i}{\mu}}{s}\right) = \frac{1}{s}\log\frac{X_i}{\mu} = \frac{1}{s}\log\left(\frac{X_i\sqrt{\bar{X}^2+\sigma^2}}{\bar{X}^2}\right) \tag{2.83}$$

For the numerical values as given, this reduces to

$$V_i = 2.5957 \cdot \log(1.077 \cdot X_i); \quad X_i = 0.92848 \cdot \exp(0.38525 V_i) \tag{2.84}$$

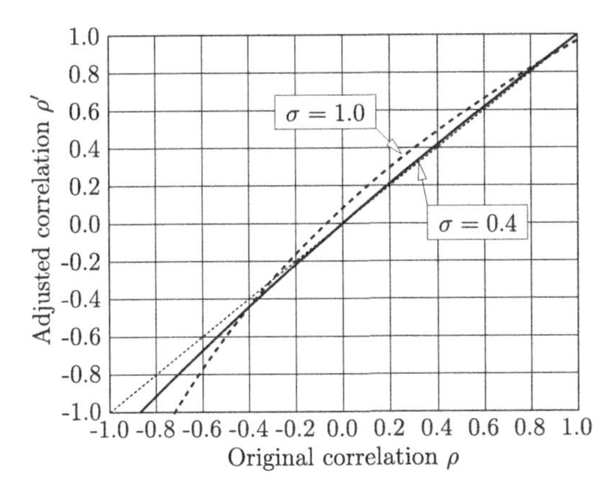

Figure 2.11 Relation between original and adjusted correlation for two correlated log-normal variables using the Nataf model.

The joint pdf can be derived as shown, e.g. in Noh, Choi, and Du 2008:

$$f_{X_1,X_2}(x_1, x_2) = \frac{1}{2\pi s_1 s_2 \sqrt{1-\rho'^2} x_1 x_2} \cdot \exp\left[-\frac{v_1^2 - 2\rho' v_1 v_2 + v_2^2}{2(1-\rho'^2)}\right] \tag{2.85}$$

in which

$$v_i = \frac{\log x_i - \log \mu_i}{s_i}; \quad i = 1, 2 \tag{2.86}$$

The coefficient of correlation ρ' in **v**-space according to Eq. 2.82 is to be determined from

$$0.16 \cdot \rho = \int_{-\infty}^{\infty}\int_{-\infty}^{\infty} (x_1 - 1)(x_2 - 1) f_{X_1X_2}(x_1, x_2) \mathrm{d}x_1\, \mathrm{d}x_2 \tag{2.87}$$

An explicit solution is given by Noh, Choi, and Du 2008

$$\rho' = \frac{\log(1 + \rho s_1 s_2)}{\sqrt{\log(1 + s_1^2)\log(1 + s_2^2)}} \tag{2.88}$$

This relation is shown in Fig. 2.11 for $\sigma = 0.4$ and $\sigma = 1$. It can be seen that the difference between ρ and ρ' is very small for positive values. However, when ρ approaches the lower limit of -1, there is no acceptable solution for ρ' leading to a positive definite correlation matrix. This effect is further increased by larger values of σ. Hence, the Nataf model ceases to function properly in this range. For a more specific example, assume a coefficient of correlation $\rho = 0.3$. In this case, the correlation coefficient in

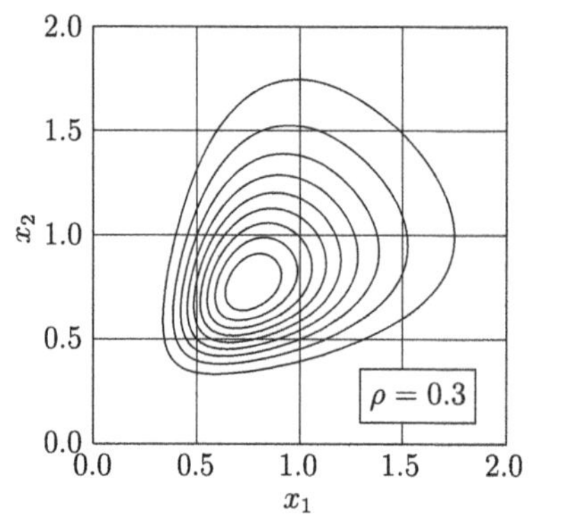

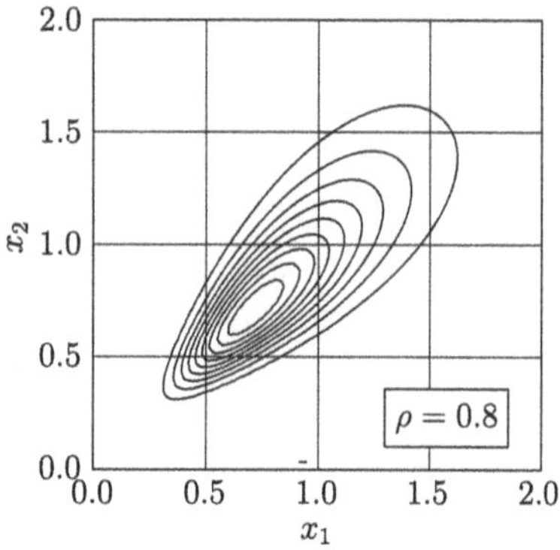

Figure 2.12 Contour lines of joint probability density function for correlated lognormally random variables based on the Nataf model.

```
1  xlo=1e-4
2  xhi=2
3  N=60
4  dx=(xhi-xlo)/(N-1)
5  s=sqrt(log(1.16))
6  rho = 0.8
7  rhop=log(1+rho*s^2)/log(1+s^2)
8  mu=exp(-s^2/2)
9  fid=fopen('joint08.txt', 'w');
10 for i=1:N
11   x1 = xlo+(i-1)*dx;
12   y1=(log(x1)-log(mu))/s;
13   for k=1:N
14     x2 = xlo+(k-1)*dx;
15     y2 = (log(x2)-log(mu))/s;
16     arg=(y1^2-2*rhop*y1*y2+y2^2)/2/(1-rhop^2);
17     deno = 2*pi*s^2*sqrt(1-rhop^2)*x1*x2;
18     pdf = 1/deno*exp(-arg);
19     fprintf(fid, '%g', pdf);
20   end
21   fprintf(fid, '\n');
22 end
23 fclose(fid);
```

Listing 2.3 Joint probability density function for two correlated lognormally distributed random variables.

Gaussian space according to Eq. 2.88 becomes $\rho' = 0.315$. For a different case with $\rho = 0.8$ we get $\rho' = 0.811$. For both cases, the contour lines of the joint probability density function are shown in Fig. 2.12. The `octave`-script generating the data for the contour plots is shown in Listing 2.3.

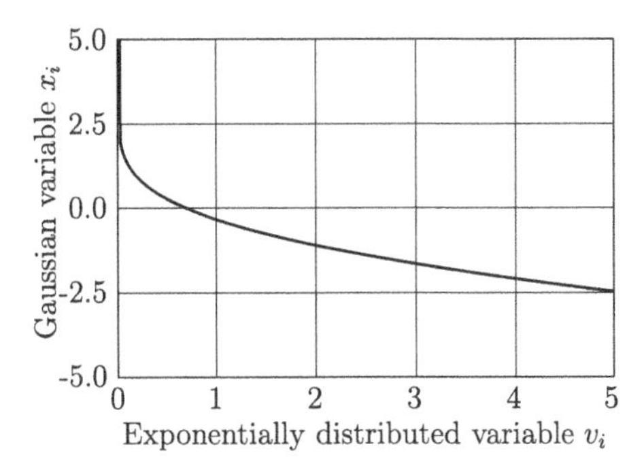

Figure 2.13 Mapping between exponential and Gaussian random variables as required for the Nataf model.

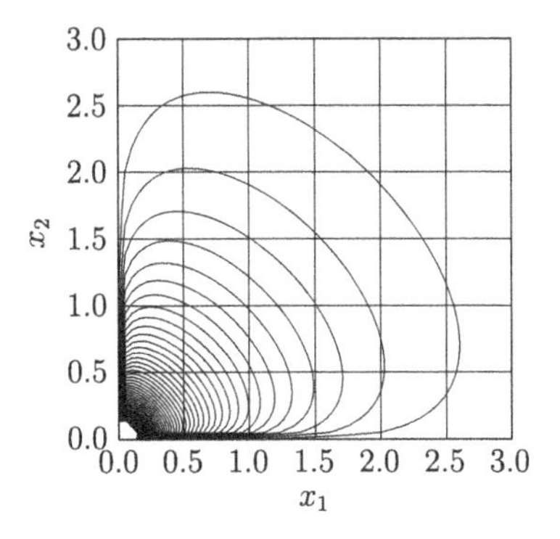

Figure 2.14 Contour lines of joint probability density function for correlated exponential random variables based on the Nataf model.

Example 2.6 (Nataf model for two exponentially distributed random variables)
Let X_1 and X_2 be two exponentially distributed random variables with mean values $\bar{X}_1 = \bar{X}_2 = 1$ and a coefficient of correlation $\rho = 0.4$. In this case, the correlation coefficient in Gaussian space becomes $\rho' = 0.4463$. The transformation from original to Gaussian space is given by

$$v_i = \Phi^{-1}[\exp(-x_i)]; x_i \geq 0 \tag{2.89}$$

This is relation is shown in Fig. 2.13. It can be observed that the transformation becomes singular as $v_i \to 0$. As a consequence, the joint probability density function of V_1 and V_2 is singular at the origin. A contour plot of this joint pdf is shown in Fig. 2.14. An detailed discussion concerning two correlated random variables with exponential marginal distributions can be found in Nadarajah and Kotz 2006.

2.2.7 *Marginal and conditional distribution*

Given the joint density function $f_{XY}(x, y)$ of two random variables, the conditional density functions are obtained by treating one of the variables as a fixed parameter. Visually, this corresponds to slicing through the "probability hill" as indicated in Fig. 2.15

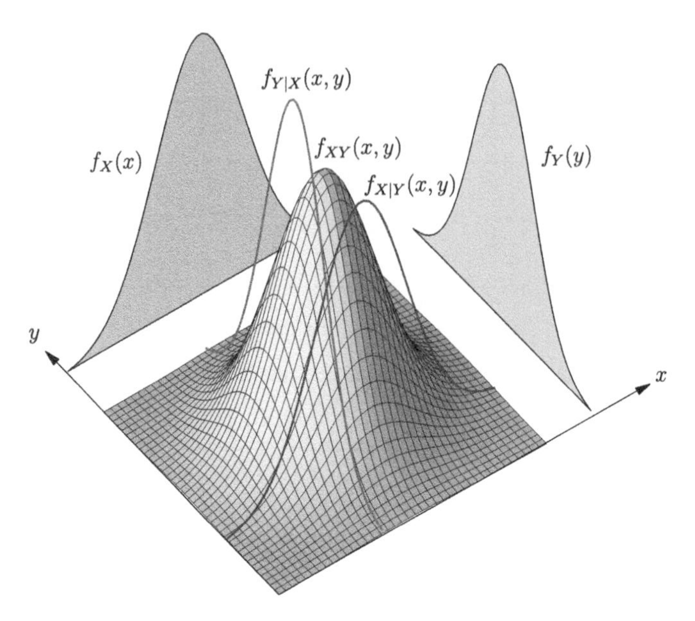

Figure 2.15 Joint density, conditional densities, and marginal densities.

According to the definition of conditional probability we have

$$f_{X|Y}(x,y) = \frac{f_{XY}(x,y)}{f_Y(y)} \tag{2.90}$$

and, by reversing the roles of X and Y

$$f_{Y|X}(x,y) = \frac{f_{XY}(x,y)}{f_X(x)} \tag{2.91}$$

The marginal density functions $f_X(x)$ and $f_Y(y)$ are obtained by integrating over one variable:

$$f_X(x) = \int_{-\infty}^{\infty} f_{XY}(x,y)\mathrm{d}y; \quad f_Y(y) = \int_{-\infty}^{\infty} f_{XY}(x,y)\mathrm{d}x \tag{2.92}$$

For jointly Gaussian random variables X and Y all conditional and marginal density functions are Gaussian.

2.3 Estimation

2.3.1 *Basic properties*

In order to connect the above mentioned concepts of probability theory to observations in reality (e.g. measurements) we need to be able to estimate statistical properties such as mean values or standard deviations from a set of available samples. In order to be useful, the estimators (which are usually implemented in terms of a specific type of algorithm) must meet certain requirements.

Assume we want to estimate a parameter γ from m independent samples $X^{(1)}, \dots X^{(m)}$ using the estimator Γ: $\Gamma_m = \Gamma(X^{(1)}, \dots X^{(m)})$.

Definition: The estimator Γ is called *consistent* if

$$\forall \epsilon > 0 : \quad \lim_{m \to \infty} \mathbf{Prob}[|\Gamma_m - \gamma| < \epsilon] = 1 \tag{2.93}$$

This means that the difference between the true value γ of the statistical parameter and its estimate Γ_m approaches zero in terms of a probability measure (i.e. convergence in probability) as the sample size m approaches infinity.

Definition: The estimator Γ is called *unbiased* if:

$$\mathbf{E}[\Gamma_m] = \gamma \tag{2.94}$$

This property implies that even in the case of finite sample size, at least "on average", we get the correct value.

Example 2.7 (Estimators for mean value and variance)
The arithmetic mean

$$M_m = \frac{1}{m} \sum_{k=1}^{m} X^{(k)} \tag{2.95}$$

is a consistent and unbiased estimator for the mean value $\mathbf{E}[X] = \bar{X}$ if the variance σ_X^2 of X is finite. Unbiasedness is readily shown as follows

$$\mathbf{E}[M_m] = \mathbf{E}\left[\frac{1}{m} \sum_{k=1}^{m} X^{(k)}\right] = \frac{1}{m} \sum_{k=1}^{m} \mathbf{E}[X^{(k)}] = \frac{1}{m} \sum_{k=1}^{m} \bar{X} = \bar{X} \tag{2.96}$$

The proof of consistency is based on Chebyshev's inequality. For this we need to compute the variance for the arithmetic mean which is

$$\begin{aligned} \sigma_{M_m}^2 &= \mathbf{E}[(M_m - \bar{X})^2] = \mathbf{E}\left[\left(\frac{1}{m} \sum_{k=1}^{m} X^{(k)} - \bar{X}\right)^2\right] \\ &= \mathbf{E}\left[\frac{1}{n^2} \sum_{k=1}^{m} \sum_{\ell=1}^{m} X^{(k)} X^{(\ell)} - \frac{2}{m} \bar{X} \sum_{k=1}^{m} X^{(k)} + \bar{X}^2\right] \\ &= \frac{1}{m^2}(m\sigma_X^2 + m^2 \bar{X}^2) - \frac{2}{m} m \bar{X}^2 + \bar{X}^2 = \frac{\sigma_X^2}{m} \end{aligned} \tag{2.97}$$

Hence, the variance of the arithmetic mean approaches zero as the sample size m approaches ∞, and from Chebyshev's inequality we have

$$\lim_{m \to \infty} \mathbf{Prob}[|M_m - \bar{X}| > \epsilon] \leq \lim_{m \to \infty} \frac{\sigma_{M_m}^2}{\epsilon^2} = \lim_{m \to \infty} \frac{\sigma_X^2}{m\epsilon^2} = 0 \qquad \forall \epsilon > 0 \tag{2.98}$$

which means that we have convergence in probability.

The sample variance

$$S_m = \frac{1}{m-1} \sum_{k=1}^{m} (X^{(k)} - M_m)^2 \tag{2.99}$$

is a consistent and unbiased estimator for σ_X^2. The latter property can be shown as follows. We begin by defining an estimator Γ_m as

$$\Gamma_m = \frac{1}{m} \sum_{k=1}^{m} (X^{(k)} - M_m)^2 \tag{2.100}$$

Taking expectation results in

$$\begin{aligned}
\mathbf{E}[\Gamma_m] &= \frac{1}{m} \sum_{k=1}^{m} \mathbf{E}[(X^{(k)} - M_m)^2] \\
&= \mathbf{E}\left[\frac{1}{m} \sum_{k=1}^{m} \left\{ X^{(k)} - \frac{1}{m} \sum_{\ell=1}^{m} X^{(\ell)} \right\}^2 \right] \\
&= \frac{1}{m} \mathbf{E}\left[\sum_{k=1}^{m} \left\{ (X^{(k)})^2 - \frac{2}{m} X^{(k)} \sum_{\ell=1}^{m} X^{(\ell)} + \frac{1}{m^2} \sum_{\ell=1}^{m} \sum_{r=1}^{m} X^{(\ell)} X^{(r)} \right\} \right] \\
&= \frac{1}{m} \left\{ \sum_{k=1}^{m} \mathbf{E}[(X^{(k)})^2] - \frac{2}{m} \sum_{k=1}^{m} \sum_{k=1}^{m} \mathbf{E}[X^{(k)} X^{(\ell)}] + \frac{1}{m^2} \sum_{k=1}^{m} \sum_{\ell=1}^{m} \sum_{r=1}^{m} \mathbf{E}[X^{(\ell)} X^{(r)}] \right\} \\
&= \frac{1}{m} \left\{ m\sigma_X^2 + m\bar{X}^2 - \frac{2}{m}(m\sigma_X^2 + m^2\bar{X}^2) + \frac{1}{m^2}(m^2\sigma_X^2 + m^3\bar{X}^2) \right\} \\
&= \frac{m-1}{m} \sigma_X^2
\end{aligned} \tag{2.101}$$

From that result it can be concluded that

$$S_m = \frac{m}{m-1} \cdot \Gamma_m = \frac{1}{m-1} \sum_{k=1}^{m} (X^{(k)} - M_m)^2 \tag{2.102}$$

is an unbiased estimator for σ_X^2.

Maximum likelihood estimation

Assume that we want to estimate the parameter γ of the probability density function $f_X(x)$ of the random variable X. This implies that we already know the type of distribution of X. For a particular value of the parameter γ, we can consider the probability density function using this parameter as a conditional probability density function $f_X(x|\gamma)$. Given m realizations $X^{(k)}, k = 1 \dots m$ of the random variable X we can compute the so-called *Likelihood Function*

$$L = \prod_{k=1}^{m} f_X(X^{(k)}|\gamma) \tag{2.103}$$

The maximum likelihood estimator of γ is chosen such as to maximize L. Frequently, it is computationally more convenient to maximize the natural logarithm of L, the so-called *Loglikelihood function* so that the condition for determining γ becomes

$$\frac{\partial}{\partial \gamma} \log L = \frac{\partial}{\partial \gamma} \sum_{k=1}^{m} \log f_X(X^{(k)}|\gamma) = 0 \tag{2.104}$$

Example 2.8 (Estimation of the parameter of an exponential distribution)
Assume X to be a random variable with the exponential probability density function $f_X(x) = \frac{1}{\lambda} \exp(-\frac{x}{\lambda})$, $x \geq 0$. Given five samples of X (i.e. 12, 14, 9, 7, 10), estimate λ using the maximum likelihood method.

The loglikelihood function is

$$\log L = -5 \log \lambda + \sum_{k=1}^{5} \left(-\frac{x^{(k)}}{\lambda} \right) \tag{2.105}$$

From this we immediately get

$$\frac{\partial}{\partial \lambda} \log L = -\frac{5}{\lambda} + \sum_{k=1}^{5} \frac{x^{(k)}}{\lambda^2} = 0 \tag{2.106}$$

with the solution

$$\lambda = \frac{1}{5} \sum_{k=1}^{5} x^{(k)} = \frac{12 + 14 + 9 + 7 + 10}{5} = 10.4 \tag{2.107}$$

It turns out that this matches the arithmetic mean of the samples.

Exercise 2.3 (Simple statistics)
Assume that repeated measurements of a physical quantity had the outcomes $X^{(k)}$ as shown in Table 2.2. Assume that these outcomes are statistically independent samples of a random variable X. Carry out the following steps:

1. Estimate the mean value $\bar{X}$ and the standard deviation σ_X from the given samples
2. Derive a linear transformation which transforms X into a random variable Y with zero mean and unit standard deviation.
3. Compute the corresponding samples of Y from the samples of X.

Table 2.2 Random samples.

k	$X^{(k)}$	$Y^{(k)}$
1	10	0.104
2	11	0.416
2	8	−1.144
4	13	1.456
5	9	−0.624

Solution: The estimated statistical parameters are $\bar{X} = 10.2$, $\sigma_X = 1.92$ and the samples of Y are given in Table 2.2 (use the relation $Y = (X - \bar{X})/\sigma_X$).

2.3.2 *Confidence intervals*

When estimating some statistical property γ of a random variable from m independent samples using an unbiased estimator Γ_m, it is helpful to characterize the precision of the estimator by providing a range in which the true value γ is enclosed with high probability so that

$$\textbf{Prob}[\Gamma_m - \epsilon \leq \gamma \leq \Gamma_m + \delta] \geq 1 - \alpha \tag{2.108}$$

for a suitable defined triple of real numbers α, δ, $\epsilon > 0$. In most cases, the level of significance α is prescribed (say at $\alpha = 0.05$ or $\alpha = 0.01$) and the quantities ϵ and δ are computed such that equality in Eq. 2.108 is satisfied. The interval

$$I_c = [\Gamma_m - \epsilon, \Gamma_m + \delta] \tag{2.109}$$

is called *confidence interval* for the confidence level $1 - \alpha$. For simplicity, we assume $\epsilon = \delta$ in the following. A first bound for the magnitude of the confidence interval can be computed from Chebyshev's inequality (2.25):

$$\textbf{Prob}[|\Gamma_m - \gamma| > \epsilon] \leq \frac{\sigma^2_{\Gamma_m}}{\epsilon^2} = \alpha \tag{2.110}$$

From this we can easily get

$$\epsilon \leq \frac{\sigma_{\Gamma_m}}{\sqrt{\alpha}} = f_C \cdot \sigma_{\Gamma_m} \tag{2.111}$$

Unfortunately, this bound is very (usually too) wide for practical purposes.

If the distribution of the unbiased estimator Γ_m (or at least its asymptotic distribution for $m \to \infty$) is known to be Gaussian, then a much narrower interval can be readily computed from the standard deviation σ_{Γ_m} of the estimator. If we assume $\epsilon = \delta$ (i.e. we are interested in deviations to larger and smaller values) then we can easily get

$$\epsilon = \delta = \Phi^{-1}\left(1 - \frac{\alpha}{2}\right) \cdot \sigma_{\Gamma_m} = f_G \cdot \sigma_{\Gamma_m} \tag{2.112}$$

Table 2.3 Width of confidence interval depending on significance level.

α	*0.1*	*0.05*	*0.02*	*0.01*
f_C	3.16	4.47	7.07	10.00
f_G	1.65	1.96	2.33	2.58

Typical values for f_C and f_G depending on α are given in Table 2.3. The choice of α depends on two factors. The first one is obviously the desire to make the probability of covering the true value within the confidence interval large. This leads to choosing small values for α. Not covering the true value by the confidence interval is called a Type II error. However, by choosing excessively small values of α we are in danger of making the interval too large, thus including too many values which are likely to be far off the true value (Type I error). Obviously, there is a trade-off between the two error possibilities involved.

Example 2.9 (Confidence interval for mean value estimator)
Consider again the arithmetic mean M_m defined in Eq. 2.95 as an estimator for the expected value of a random variable X. It is unbiased and its standard deviation depends on the standard deviation σ_X of the random variable and on the number of samples m, i.e. $\sigma_{M_m} = \frac{\sigma_X}{m}$. One interesting question in statistics is, how large to choose the sample size m in order to obtain the mean within an interval of $\pm 10\%$ about the true value with a confidence level of 95%. Presuming an asymptotically Gaussian distribution for the estimator, we can compute the required sample size if we know the standard deviation of the random variable X. For this example, we assume that $\sigma_X = 0.2 \cdot \bar{X}$. Then we have

$$\epsilon = f_G \cdot \frac{\sigma_X}{\sqrt{m}} \leq 0.1\bar{X} \rightarrow m \geq \left(\frac{f_G \cdot 0.2\bar{X}}{0.1\bar{X}} \right)^2 = \left(\frac{1.96 \cdot 0.2}{0.1} \right)^2 = 15.37 \qquad (2.113)$$

in which f_G from Table 2.3 with $\alpha = 0.95$ has been used. If we are using the wider bound as obtained from Chebyshev's inequality, we get

$$m \geq \left(\frac{f_C \cdot 0.2\bar{X}}{0.1\bar{X}} \right)^2 = 79.92 \qquad (2.114)$$

2.3.3 *Chi-square test*

In many cases, we have some prior knowledge (e.g. from experience based on many prior observations) about the probability distribution of a random variable X. Let this prior knowledge be described in terms of the hypothetical distribution function $F_0(x)$. Furthermore, assume that we have samples $X^{(k)}, k = 1 \ldots m$ of the random variable X.

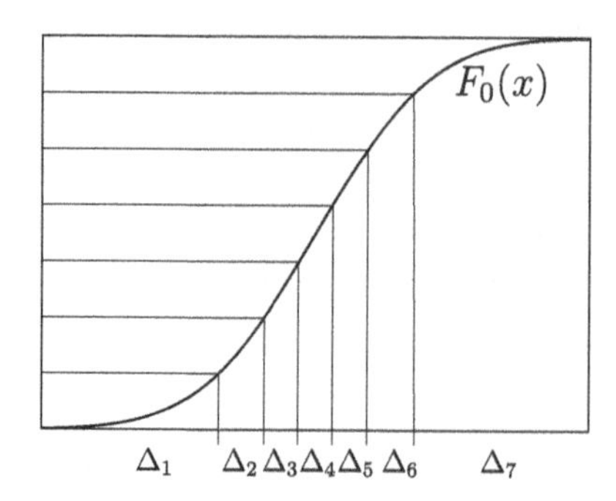

Figure 2.16 Class definition for χ^2-test.

Let the true distribution function of X be denoted as $F_X(x)$. We would like to test the hypothesis

$$H_0 : F_X(x) \equiv F_0(x) \tag{2.115}$$

by means of the available samples using the chi-square test (Bronstein and Semendjajev 1983).

In order to carry out the χ^2-Test, the range of the variable X is divided into a finite number of disjoint sets (intervals) $\Delta_j; j = 1 \ldots n_c$. The probabilities of a random sample lying in the interval Δ_j as computed from $F_0(x)$ is p_j. Here, the interval sizes should be chosen such that the the product $m \cdot p_j \geq 5$. Then the sample data are collected into the intervals Δ_j. The number of samples within each interval is denoted by m_j. Obviously, we have the relations

$$\sum_{j=1}^{n_c} p_j = 1; \quad \sum_{j=1}^{n_c} m_j = m \tag{2.116}$$

Then we compute the test statistic

$$\chi^2 = \sum_{j=1}^{n_c} \frac{(m_j - m \cdot p_j)^2}{m \cdot p_j} \tag{2.117}$$

If the hypothesis H_0 is true, then the test quantity is asymptotically χ^2-distributed with $N = n_c - 1$ degrees of freedom. Hence the hypothesis H_0 should be rejected if the test value χ^2 exceeds a certain critical value. The critical value χ^2_α depends on the significance level α as well as on the degree of freedom. This can be seen from Fig. 2.17.

Example 2.10 (Test of uniformity)
Let the outcomes of a rolling dice experiment with 60 trials be as shown in Table 2.4. We want to test the hypothesis that the outcomes are uniformly distributed (the game of dice is fair). Here that classes Δ_j are simply defined in terms of the possible outcomes. The degree of freedom is $N = 6 - 1 = 5$. For a significance level $\alpha = 0.05$ we get $\chi^2_a = 11.07$. The test value is much smaller, hence we accept the hypothesis of a fair game. The `octave`-command to get the value of χ^2_α is (note that $1 - \alpha$ and N are passed as parameters)

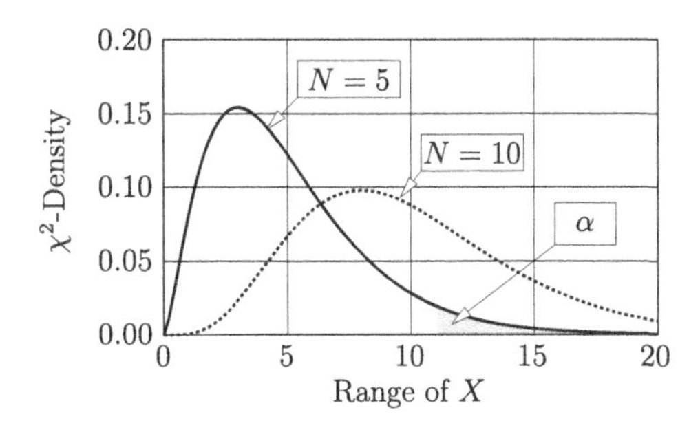

Figure 2.17 χ^2 probability density function.

Table 2.4 Test of uniformity for dice experiment.

j	m_j	$m{\cdot}p_j$	$\frac{(m_j - m{\cdot}p_j)^2}{m{\cdot}p_j}$
1	9	10	0.1
2	11	10	0.1
3	12	10	0.4
4	13	10	0.9
5	8	10	0.4
6	7	10	0.9
$\sum$	60	60	2.8

```
1 chi_a = chisquare_inv(0.95, 5)
```

Distribution parameters estimated from the same data set

Frequently, the hypothesis H_0 needs to be based on the same sample which is used for the test. A typical situation is that the type of distribution is known, but the describing parameters $\theta_\ell, \ell = 1 \ldots n_\theta$ (e.g. mean value and standard deviation) have to be estimated from the sample. This case can be incorporated into the χ^2-test. In the first step, the parameters θ_ℓ are estimated using maximum likelihood estimators $\hat{\theta}_\ell$. Based on these parameters, the hypothetical distribution is $F_0(x) = F_0(x, \hat{\theta}_1, \ldots \hat{\theta}_{n_\theta})$. The classification the proceeds as described above. The critical value χ^2_α is now computed from a χ^2-distribution with a smaller degree of freedom, i.e. $N = n_c - 1 - n_\theta$.

Exercise 2.4 (Test of exponential distribution)
Let the hypothesis to be tested be

$$F_0(x, \lambda) = 1 - \exp\left(-\frac{x}{\lambda}\right) \tag{2.118}$$

We assume to have a data set of 50 samples as given in Table 2.5.

Table 2.5 Data set for test of exponential distribution

4.00	0.46	1.24	0.21	0.41
1.73	0.23	4.35	0.87	0.12
1.76	0.96	3.46	1.21	2.20
0.69	0.49	0.28	0.23	1.02
1.30	0.49	0.67	0.07	0.55
0.42	0.76	0.34	1.28	0.06
0.22	0.41	0.16	0.04	1.22
0.30	0.06	0.93	1.58	0.23
1.57	0.40	0.03	0.71	0.34
1.69	0.31	0.11	2.27	6.43

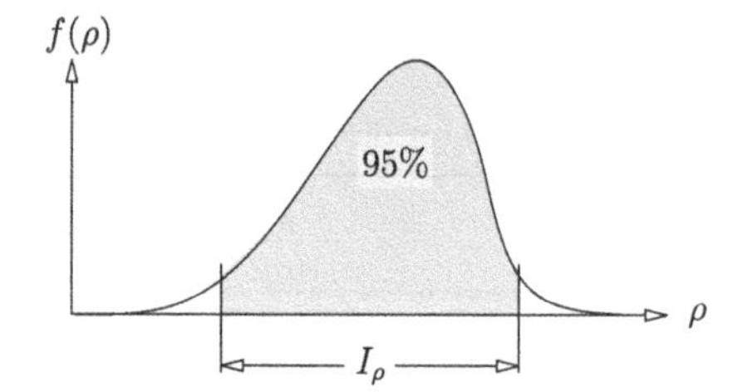

Figure 2.18 Confidence interval for estimated coefficients of correlation ρ.

Test the null hypothesis using 8 classes with widths chosen such that the theoretical probabilities within each class are identical (i.e. $p_j = \frac{1}{8}, j = 1 \ldots 8$).

Solution: In the first step we need to estimate the distribution parameter λ. The maximum likelihood estimator is $\lambda = 1.017$. We then compute the classes defined by the intervals $[c_{j-1}, c_j]$ in which $c_j = -\lambda \log(1 - \frac{j}{8})$, $j = 1 \ldots 8$. Thus we get a value of $\chi^2 = 3.12$. The degree of freedom is 6, hence the critical value is $\chi_a^2 = 12.59$ for $\alpha = 0.05$. As $\chi^2 < \chi_a^2$, there is no reason to reject null hypothesis.

2.3.4 *Correlation statistics*

Assume that we want to estimate a matrix of correlation coefficients of m variables from N samples. This matrix has $M = m \cdot (m-1)/2$ different entries in addition to m unit elements on the main diagonal. The confidence intervals for the estimated coefficients of correlation ρ_{ij} are computed based on Fisher's z-transformation. The interval for a significance level of α (i.e. a confidence level of $1 - \alpha$) is given by

$$\left[\tanh\left(z_{ij} - \frac{z_c}{\sqrt{N-3}}\right), \tanh\left(z_{ij} + \frac{z_c}{\sqrt{N-3}}\right)\right] \tag{2.119}$$

In this equation, N is the number of samples used for the estimation of ρ_{ij}. The critical value z_c is computed by using the Bonferroni-corrected value for the significance level $\alpha' = \alpha/M$ with M being the number of correlation coefficients to be estimated (see above). The transformed variable z is computed from

$$z_{ij} = \frac{1}{2} \log \frac{1 + \rho_{ij}}{1 - \rho_{ij}} \tag{2.120}$$

and the critical value z_c is given by

$$z_c = \Phi^{-1}(1 - \alpha'/2) \tag{2.121}$$

where $\Phi^{-1}(.)$ is the inverse cumulative Gaussian distribution function.

2.3.5 *Bayesian updating*

In many areas of science and engineering predictions about the behavior of systems or structures rely on mathematical models involving a number of uncertain parameters. Frequently, we are facing the situation where a set of statistical parameters $\theta_1, \ldots \theta_\ell$ describing the probability distribution $f_{X|\theta_1,\ldots\theta_\ell}(x) = f_X(x, \theta_1, \ldots \theta_\ell)$ of a random variable X is known a priori in terms of a probability distribution $f_{\Theta_1,\ldots\Theta_\ell}(\theta_1, \ldots \theta_\ell)$. This will occur, e.g., if the set of parameters has been estimated from a number of independent samples of X and the process involves some sampling uncertainty (cf. Viertl 2003).

For reasons of clarity, we will focus on the case of one single parameter Θ in the subsequent developments. The theorem of total probability can be utilized to construct the (prior) probability density function of X, i.e.

$$f_X(x) = \int_{-\infty}^{\infty} f_{X|\Theta=\vartheta}(x) f_\Theta(\vartheta) \mathrm{d}\vartheta = \int_{-\infty}^{\infty} f_X(x, \vartheta) f_\Theta(\vartheta) \mathrm{d}\vartheta \tag{2.122}$$

Assume now that we have made additional measurements $X^{(k)}, k = 1 \ldots m$ which allows us to perform a new estimate of the parameter set, say $\theta'_1, \ldots \theta'_\ell$. The question now is, how should we change the probability distribution of those parameters and, furthermore, what is the consequence on the probability distribution of X. For the following developments we need to utilize the continuous version of Bayes' theorem (Papoulis 1984):

$$f_{\Theta|\mathcal{A}}(\vartheta) = \frac{\mathbf{Prob}[\mathcal{A}|\Theta = \vartheta]}{\mathbf{Prob}[\mathcal{A}]} f_\Theta(\vartheta) = \frac{\mathbf{Prob}[\mathcal{A}|\Theta = \vartheta] \cdot f_\Theta(\vartheta)}{\int_{-\infty}^{\infty} \mathbf{Prob}[\mathcal{A}|\Theta = \vartheta] \cdot f_\Theta(\vartheta) \vartheta} \tag{2.123}$$

In our case, the event $\mathcal{A}$ is the occurrence of an observed value Θ' in the interval $[\vartheta, \vartheta + \mathrm{d}\vartheta]$ so that

$$\begin{aligned} f_{\Theta''}(\vartheta) &= f_{\Theta|\vartheta \leq \Theta' \leq \vartheta + \mathrm{d}\vartheta}(\vartheta) = \frac{\mathbf{Prob}[\vartheta \leq \Theta' \leq \vartheta + \mathrm{d}\vartheta|\Theta = \vartheta]}{\mathbf{Prob}[\vartheta \leq \Theta' \leq \vartheta + \mathrm{d}\vartheta]} f_\Theta(\vartheta) \\ &= \frac{f_{\Theta'|\Theta=\vartheta}(\vartheta) \mathrm{d}\vartheta \cdot f_\Theta(\vartheta)}{\int_{-\infty}^{\infty} f_{\Theta'|\Theta=\vartheta}(\vartheta) \mathrm{d}\vartheta \cdot f_\Theta(\vartheta) \mathrm{d}\vartheta} = \frac{f_{\Theta'|\Theta=\vartheta}(\vartheta) \cdot f_\Theta(\vartheta)}{\int_{-\infty}^{\infty} f_{\Theta'|\Theta=\vartheta}(\vartheta) \cdot f_\Theta(\vartheta) \mathrm{d}\vartheta} \end{aligned} \tag{2.124}$$

The function $f_{\Theta'|\Theta=\vartheta}(\vartheta)$ is the likelihood of the occurrence of an observation Θ' given the prior information that $\Theta = \vartheta$. $f_{\Theta''}(\vartheta)$ denotes the so-called *posterior* probability density function of the parameter ϑ.

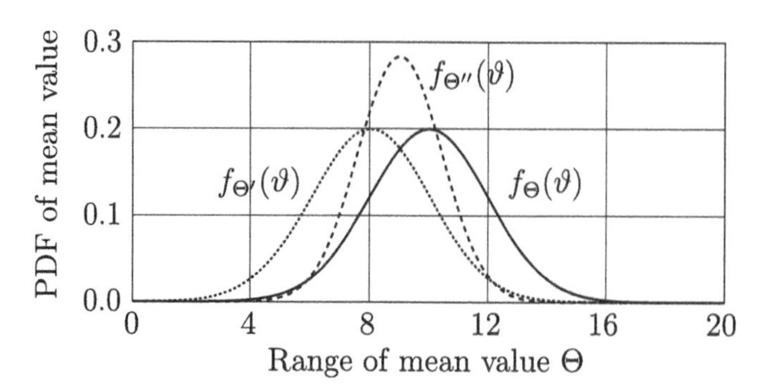

Figure 2.19 Bayesian updating of the probability density of the mean value.

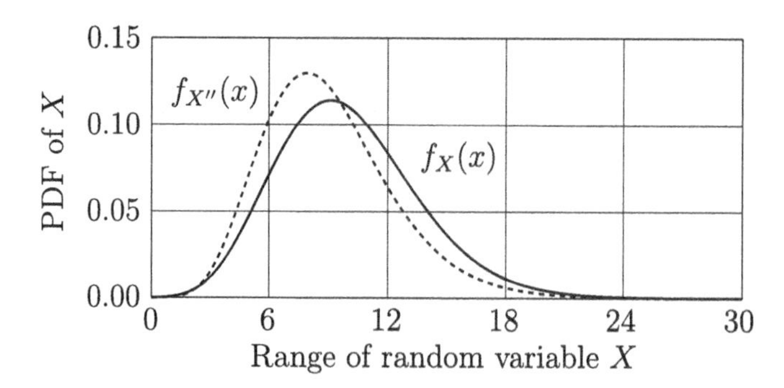

Figure 2.20 Bayesian predictive probability density of a log-normal random variable.

In the next step, we construct the so-called *predictive probability density* for the random variable X which takes into account the updated probability density for the parameter set. This, again, is based on the total probability theorem resulting in

$$f_{X''}(x) = \int_{-\infty}^{\infty} f_{X|\Theta=\vartheta}(x) f_{\Theta''}(\vartheta) \mathrm{d}\vartheta = \int_{-\infty}^{\infty} f_X(x,\vartheta) f_{\Theta''}(\vartheta) \mathrm{d}\vartheta \tag{2.125}$$

Example 2.11 (Updating of a Log-normal distribution)
Assume that we know the distribution of a random variable X to be log-normal. We are given the prior information that its standard deviation is $\sigma_X = 3$ and that its expected value $\Theta = \bar{X}$ is a random variable with a mean value of $\bar{\Theta} = 10$ and a standard deviation of $\sigma_\Theta = 2$ (cf. Fig. 2.19). Now we carry out three independent measurements with $X^{(1)} = 8.0$, $X^{(2)} = 7.2$, and $X^{(3)} = 8.9$. From this we estimate $E[\Theta'] = 8.03$.

We carry out the updating for Θ as given by Eq. 2.124. Using numerical integration, we obtain the posterior distribution of Θ as shown in Fig. 2.19. Fig. 2.20 shows that the probability density function of X shifts towards smaller values due to the updating process. This is in agreement with the shift of the probability density of the mean value towards smaller values as shown in Fig. 2.19. The entire updating process is given in the `octave`-code shown in Listing 2.4.

```
1  samples=[8,7.2,8.9];
2  m=mean(samples);
3  s=std(samples);
4
5  sigma = 3;
6
7  nt=200;
8  tu=0.001;
9  to=20;
10 dt = (to-tu)/nt;
11 t=linspace(tu, to,nt);
12
13 nx=200;
14 xu=0;
15 xo=30;
16 dx=(xo-xu)/nx;
17 x=linspace(xu,xo,nx);
18
19 fx=normpdf(t, 10, 2);
20 fy=normpdf(t, m, 2);
21 fxfy=fx.*fy;
22 sum=sum(fxfy)*dt;
23 fxfy=fxfy/sum;
24 plot(t,fx,t,fy,t,fxfy)
25 pause
26
27 fnew=zeros(1,nx);
28 fold=zeros(1,nx);
29 for i=1:nt
30  theta=t(i);
31   mu = log(theta ^2/sqrt(theta ^2+sigma ^2));
32   s = sqrt(log(sigma ^2/theta ^2 + 1));
33   help=lognpdf(x,mu,s)*fx(i)*dt;
34   fold = fold + help;
35   help=lognpdf(x,mu,s)*fxfy(i)*dt;
36   fnew = fnew + help;
37 end
38 plot(x,fold,x,fnew)
39 pause
```

Listing 2.4 Bayesian updating.

Exercise 2.5 (Updating of an exponential distribution)
Consider a case in which the statistical parameter A under consideration is the mean value of a random variable and is exponentially distributed with mean value 1.

Now we have 4 additional measurements of X, i.e. $X^{(1)}=1.1$, $X^{(2)}=1.2$, $X^{(3)}=0.9$, and $X^{(4)}=1.3$. Perform a Bayesian updating of the distribution parameter and compute $f_{A''}(a)$.

Solution: The updated probability density function of the parameter A is

$$f_{A''}(a) = \frac{5.0915}{a} \exp\left(-\frac{a^2+1.125}{a}\right) \tag{2.126}$$

2.3.6 *Entropy concepts*

The term *entropy* in the context of probability theory refers to the degree of uncertainty associated with a random variable. In that sense, it is a measure of (missing) information. Formally, the entropy H of a random variable X (associated with a probability density function $f_X(x)$) is defined as (Papoulis 1984)

$$H(X) = -\int_C^D f_X(x) \log f_X(x) \mathrm{d}x \tag{2.127}$$

in which the limits of integration C and D have to be chosen so that they represent the support of the density function, i.e. so that $f_X(x) \neq 0, \forall x \in [C, D]$. As an example, consider a Gaussian random variable X with mean value $\bar{X}$ and standard deviation σ_X. Using the substitution $u = \frac{x-\bar{X}}{\sigma_X}$ its entropy is easily computed as

$$\begin{aligned}
H(X) &= -\int_{-\infty}^{\infty} \sigma_X f_X(u) \log f_X(u) \mathrm{d}u \\
&= -\int_{-\infty}^{\infty} \frac{\sigma_X}{\sqrt{2\pi}\sigma_X} \exp\left(-\frac{u^2}{2}\right)\left[\log\frac{1}{\sqrt{2\pi}\sigma_X} - \frac{u^2}{2}\right] \mathrm{d}u \\
&= \frac{1}{\sqrt{2\pi}}\left[\log\sqrt{2\pi}\sigma_X \int_{-\infty}^{\infty} \exp\left(-\frac{u^2}{2}\right) \mathrm{d}u + \frac{1}{2}\int_{-\infty}^{\infty} u^2 \exp\left(-\frac{u^2}{2}\right) \mathrm{d}u\right] \\
&= \log\sqrt{2\pi}\sigma_X + \frac{1}{2}
\end{aligned} \tag{2.128}$$

This shows that the entropy H of a Gaussian random variable X increases with the standard deviation σ_X but does not depend on the mean value $\bar{X}$.

In the context of probabilistic modeling, the so-called maximum entropy method makes use of the idea that by maximizing the entropy subject to the available statistical information we leave as much uncertainty as possible without introducing subjective information. A thorough discussion of information theory in a statistical context can be found in Kullback 1997.

A typical application is the case that we know some expected values $\mathbf{E}[g_\ell(X)]$ of known functions $g_\ell(.), \ell = 1 \ldots m$ of the random variable X (such as, e.g., statistical moments or centralized statistical moments, cf. Eq. 2.21) and that we want to infer the probability density function $f_X(x)$ using this information only. As shown in Papoulis 1984, this leads to a probability density function in exponential form

$$f_X(x) = A \exp\left[-\sum_{\ell=1}^{m} \lambda_\ell g_\ell(x) \right] \tag{2.129}$$

in which A is a normalizing constant and $\lambda_\ell, \ell = 1, m$ are Lagrange multipliers to be determined in a way that the given expectations are satisfied:

$$\begin{aligned} E[g_\ell(X)] &= \int_{-\infty}^{\infty} g_\ell(x) f_X(x) \mathrm{d}x \\ &= \int_{-\infty}^{\infty} g_\ell(x) A \exp\left[-\sum_{\ell=1}^{m} \lambda_\ell g_\ell(x) \right] \mathrm{d}x; \; \ell = 1, m \end{aligned} \tag{2.130}$$

In most cases, these equations cannot be solved in closed form so that a numerical scheme is required. It should be noted that in the case of given statistical moments, i.e. $g_\ell(x) = x^\ell$; $\ell = 1 \ldots m$, it is important to be aware of the fact that either m must be even, or that the support of the random variable is not extending to both $-\infty$ and $+\infty$. Otherwise, the integrals in Eq. 2.130 will not exist.

Example 2.12 (Maximum entropy distribution given the mean value)
Consider the case of a random variable X which is known to be positive. We also know the mean value $\bar{X}$. From this information we want to construct the maximum entropy probability density function $f_X(x)$. Here, the prescribed expected value is $\mathbf{E}[g_1(X)] = \mathbf{E}[X]$

$$\mathbf{E}[X] = \int_{0}^{\infty} g_1(x) f_X(x) \mathrm{d}x = \int_{0}^{\infty} x f_X(x) \mathrm{d}x \tag{2.131}$$

and the probability density function is

$$f_X(x) = A \exp(-\lambda_1 g_1(x)) = A \exp(-\lambda x) \tag{2.132}$$

From the normalization condition (i.e. from the condition that the probability density function integrates to 1) we get

$$1 = \int_{0}^{\infty} f_X(x) dx = \int_{0}^{\infty} A \exp(-\lambda x) \mathrm{d}x = \frac{A}{\lambda} \rightarrow A = \lambda \tag{2.133}$$

From the mean value condition we get

$$\bar{X} = \int_0^\infty x f_X(x) \mathrm{d}x = \lambda \int_0^\infty x \exp(-\lambda x) \mathrm{d}x = \frac{1}{\lambda} \rightarrow \lambda = \frac{1}{\bar{X}} \tag{2.134}$$

So we get an exponential distribution with the density function

$$f_X(X) = \frac{1}{\bar{X}} \exp\left(-\frac{x}{\bar{X}}\right) \tag{2.135}$$

Exercise 2.6 (Maximum entropy given the mean and the standard deviation)
Consider the case of an unbounded random variable X with given mean value $\bar{X}$ and standard deviation σ_X. Show that the maximum entropy distribution is a Gaussian with given mean and standard deviation.

2.4 Simulation techniques

2.4.1 *General remarks*

As already demonstrated in chapter 1, there are many problems in stochastic structural analysis which can be treated by means of random number generators in combination with statistical methods. In order to provide some background information on the available methods, this section deals with some methods to generate samples of random variables and random vectors by means of digital computation. The purpose of the entire process is to obtain probabilistic information through the application of statistical methods to these samples. In order to fulfill the requirements, the samples need to follow the prescribed distribution functions closely. An additional requirement is that the generated samples are essentially statistically independent of each other. This is a prerequisite for the applicability of the available statistical methods. Finally, statistical methods are utilized to estimate expected values or probabilities of the structural response quantities of interest.

Typically, the analysis process is then structured according to the flow chart as given in Fig. 2.21.

From a mathematical perspective, the process of computing expected values is equivalent to an integration process. Hence sample-based simulation techniques may be viewed as specific types of integration methods. One of the major problems encountered in classical numerical integration methods (such as e.g. Gauss-integration) is the so-called curse of dimensionality, i.e. the number of integration points required grows exponentially with the number of variables. Such a growth cannot be accepted for high-dimensional problems involving several hundreds of random variables.

2.4.2 *Crude Monte Carlo simulation*

This is a frequently used method to deal with the effect of random uncertainties. Typically its application aims at integrations such as the computation of expected values (e.g. mean or standard deviation).

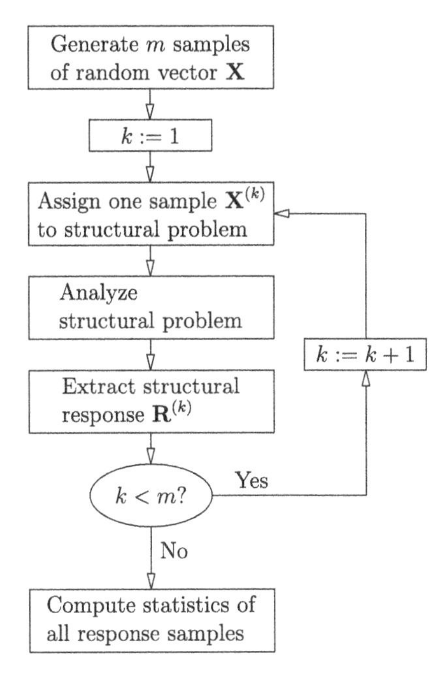

Figure 2.21 Flow of computation for sample-based stochastic structural analysis.

In order to illustrate the close relationship between the computation of probabilities and integration, consider the determination of the area of a quarter circle of unit radius. As we know, the area is $\frac{\pi}{4}$, which can be computed using analytical integration. Using the Monte Carlo Method we can obtain approximations to this result based on elementary function evaluations. When we use 1000 uniformly distributed random numbers x and y (cf. Fig. 2.22), and count the number N_c of pairs (x, y) for which $x^2 + y^2 < 1$, we get an estimate $\frac{\pi}{4} \approx \frac{N_c}{1000} = \frac{791}{1000} = 0.791$. This differs from the exact result $\frac{\pi}{4} = 0.7854$ by about 1%. The `octave`-script generating the data is given in Listing 2.5.

Running this script we obtain

```
1  M =   1000
2  NC = 791
3  area =   0.79100
```

Of course, the result will be different in each run and will be stabilized by increasing the number of samples.

2.4.3 *Latin Hypercube sampling*

In order to reduce the statistical uncertainty associated with Monte Carlo estimation of expected values, alternative methods have been developed. One such strategy is

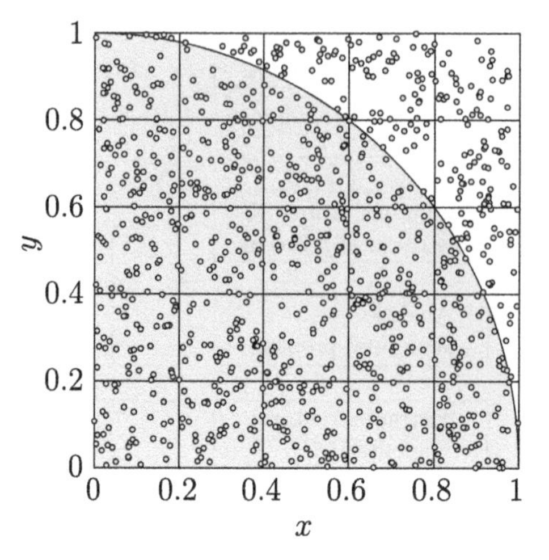

Figure 2.22 Estimating $\frac{\pi}{4}$ by Monte Carlo simulation.

```
1  M=1000
2  x=rand(M,1);
3  y=rand(M,1);
4  r2=x.^2+y.^2;
5  indic = r2<1;
6  NC=sum(indic)
7  area=NC/M
8  fid=fopen('circle.txt','w');
9  for i=1:size(x)
10   fprintf(fid, '%g %g\n', x(i), y(i));
11 end
12 fclose(fid);
```

Listing 2.5 Estimating $\frac{\pi}{4}$ by Monte Carlo simulation.

the Latin Hypercube sampling method (Imam and Conover 1982). Essentially, this method aims at ensuring a good coverage of the random variable space. For simplicity, we discuss the procedure as formulated in the space of independently and identically uniformly distributed random variables $X_i, i = 1 \ldots n$. We assume that m samples of these random variables are to be generated. In the first step, the range of the variables is divided into m intervals of equal size $\frac{1}{m}$. Then one value in each interval is chosen as a representative value, e.g. the value in the center of the interval. For the one-dimensional case $n = 1$, the sampling procedure simply consists of drawing these representative values one by one in random order. In this way, the range of the variable is automatically covered. For higher-dimensional cases $n > 1$, however, the situation becomes more complex. This is due to fact that covering all possible combinations leads to an exponential growth of the number of required samples. Hence the restrictions on

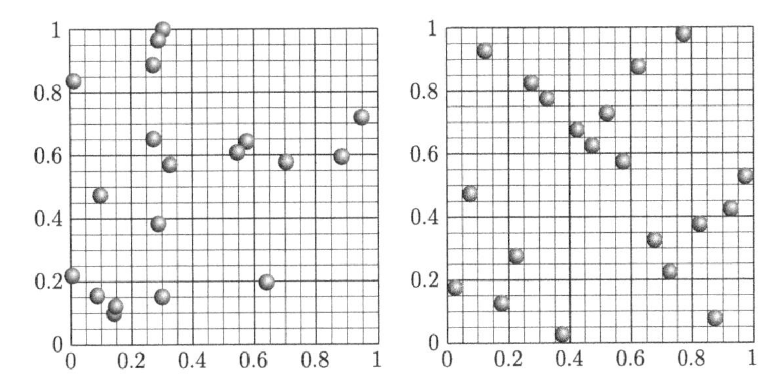

Figure 2.23 Crude Monte Carlo (left) vs. Latin hypercube (right) sampling.

Table 2.6 95% confidence interval of correlation coefficient, plain Monte Carlo.

N	*ρ*				
	0	*0.3*	*0.5*	*0.7*	*0.9*
10	1.261	1.231	1.054	0.757	0.299
30	0.712	0.682	0.557	0.381	0.149
100	0.409	0.374	0.306	0.199	0.079
300	0.230	0.209	0.170	0.116	0.045
1000	0.124	0.115	0.093	0.062	0.023

the computational effort require keeping the number of samples small. However, in this case the random drawing may introduce unwanted patterns in the sample distribution which may lead to undesired correlations between the random variables. A strategy to reduce this correlation has been presented in Florian 1992. Fig. 2.23 compares the samples obtained from crude Monte Carlo sampling with those obtained by Latin Hypercube sampling with reduction of the correlations. The coefficient of correlation in the left figure is $\rho = 0.272$, whereas the coefficient of correlation in the right figure is only $\rho = -0.008$, i.e., it is virtually zero. Although the correlation is significantly improved, it remains be noticed that the samples do not really fill the random variable space evenly.

In order to obtain meaningful correlations between the input and output (i.e. the structural response) variables, it is essential to precisely capture the input correlations in the simulated values. For demonstration purposes, a numerical study performing a comparison of the estimation errors of the correlation coefficients expressed in terms of the confidence interval, is carried out. Samples with a prescribed coefficient of correlation ρ are generated using both crude Monte Carlo sampling and Latin hypercube sampling, then the correlation coefficient is statistically estimated from these samples. The simulation and statistical analysis is repeated 1000 times. Tables 2.6 and 2.7 show the confidence intervals for the correlation coefficients for a confidence level of 95% as a function of the correlation coefficient ρ and the number of samples N used for each estimation obtained from crude Monte Carlo and Latin Hypercube sampling respectively.

Table 2.7 95% confidence interval of correlation coefficient, Latin Hypercube sampling.

N	ρ				
	0	*0.3*	*0.5*	*0.7*	*0.9*
10	0.420	0.382	0.260	0.158	0.035
30	0.197	0.194	0.139	0.073	0.018
100	0.111	0.101	0.071	0.042	0.009
300	0.065	0.057	0.042	0.024	0.006
1000	0.038	0.033	0.025	0.014	0.003

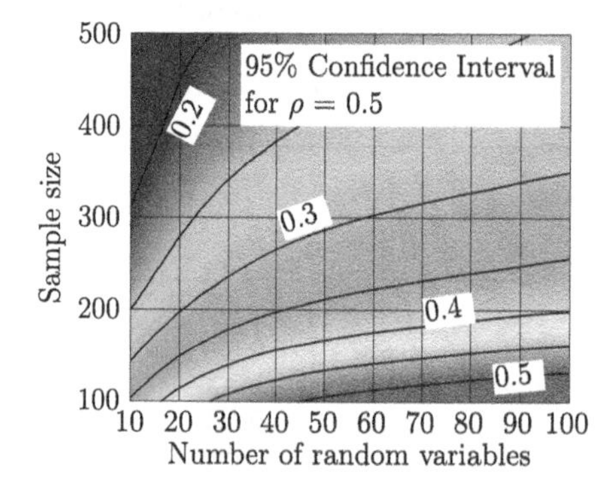

Figure 2.24 Confidence intervals for coefficients of correlation.

In summary, it turns out that the net effect of LHS is an effective reduction of the sample size by a factor of more than 10. For example, as shown in Tables 2.6 and 2.7, it is possible to estimate a coefficient of correlation of $\rho = 0.3$ using 1000 samples of MCS with a 95%-confidence interval of 0.11, while the same confidence interval (actually 0.1) is achieved with only 100 samples using LHS. On the other hand, 1000 LHS samples would reduce the respective 95%-confidence interval to 0.03, which is an enormous improvement.

2.4.4 *Quasirandom sequences*

As shown above, Latin Hypercube samples are quite suitable to represent the correlation structure of the random variables but they are not necessarily very good in covering the entire range of the variables for large dimensionality. Several approaches to generate samples have been developed with the purpose to provide better properties in covering the range of the variables. As a consequence, they reduce the estimation errors (or, in the mathematical interpretation, the integration error). An extensive discussion of such quasi-random sequences can be found in Niederreiter 1992. The main purpose of these sequences is an improved space filling which eventually leads to smaller errors in the integration results. For instance, Sobol sequences (Sobol and Asotsky 2003) can be utilized. Algorithms for generating Sobol seqences are described,

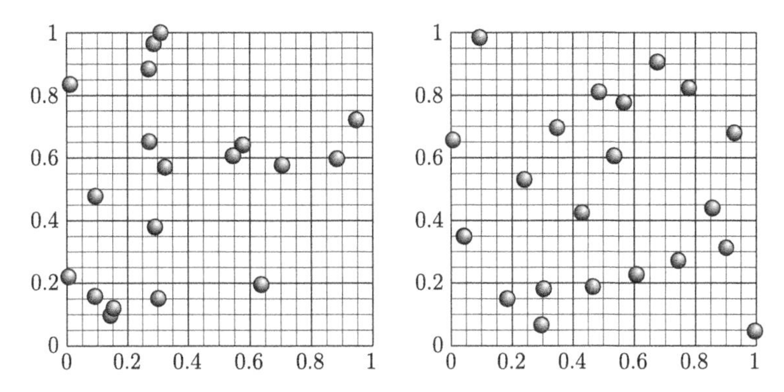

Figure 2.25 Crude Monte Carlo (left) vs. Sobol (right) sampling.

Table 2.8 Estimation quality for crude Monte Carlo (MC), Latin Hypercube (LH), and Randomized Sobol (RS) samples.

m	*MC*		*LH*		*RS*	
	A	σ_A	*A*	σ_A	*A*	σ_A
10	0.785	0.129	0.786	0.073	0.784	0.086
20	0.787	0.092	0.789	0.050	0.787	0.048
50	0.786	0.056	0.784	0.032	0.783	0.025

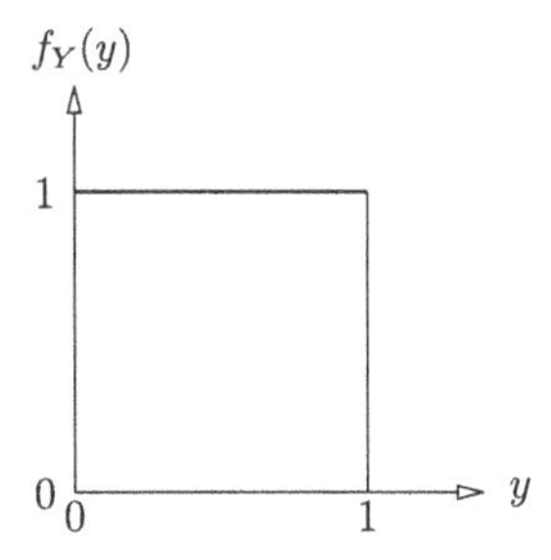

Figure 2.26 Probability density function of uniformly distributed random variable.

e.g., by Bratley and Fox 1988 and for a randomized version allowing statistical error estimates, by Hong and Hickernell 2003.

As a final example, consider again the determination of $\frac{\pi}{4}$ by sampling. Fixing the number of samples of X and Y at a value m, computing the area estimate A for these samples and repeating the analysis 1000 times we can get a measure for the estimation error in terms of the standard deviation σ_A. Table 2.8 shows the mean value $\bar{A}$ and the standard deviation σ_A for crude Monte Carlo, Latin Hypercube, and randomized Sobol samples respectively. It can be seen that the performance of Latin Hypercube and randomized Sobol samples is comparably good in this case and that crude Monte Carlo is clearly worse.

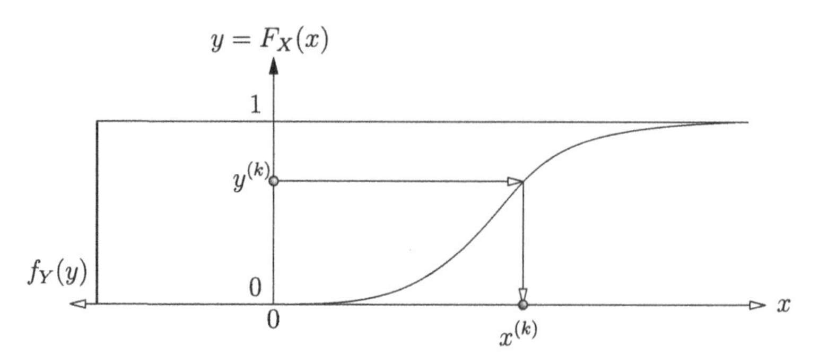

Figure 2.27 Digital generation of samples of random variable.

2.4.5 *Transformation of random samples*

An interesting application of Eq. 2.59 is the digital generation of realizations of a random variable X with prescribed distribution function $F_X(x)$ from samples of a random variable Y which is uniformly distributed in $(0, 1)$. This means that the probability density function of Y is

$$f_Y(y) = \begin{cases} 1 & \text{for } 0 \leq y \leq 1 \\ 0 & \text{else} \end{cases} \tag{2.136}$$

so that the transformation between x and y space is governed by the differential equation

$$1 = f_X(x)\frac{dx}{dy} \rightarrow \frac{dy}{dx} = f_X(x) \tag{2.137}$$

which is readily solved and inverted:

$$y = F_X(x) \rightarrow x = F_X^{-1}(y) \tag{2.138}$$

So samples $y^{(k)}$ are digitally generated and transformed into samples $x^{(k)}$ by means of Eq. 2.138.

2.4.6 *Simulation of correlated variables*

Widely used random number generators are optimized at producing sequences of number which appear to be uncorrelated. Hence the simulation of correlated random variables requires suitable transformations. The details of the transformation depend on the joint probability density function of these variables. If the joint density function is based on the Nataf model (cf. section 2.2.6), then the marginal density function $f_{X_i}(x_i), i = 1, \ldots n$ and the correlation coefficients $\rho_{ij}, i, j = 1, \ldots n$ have to be known. The simulation can then be performed in a loop for $k = 1, \ldots m$ using these steps:

1. Generate one sample $\mathbf{u}^{(k)}$ of a vector of n uncorrelated standardized Gaussian random variables.

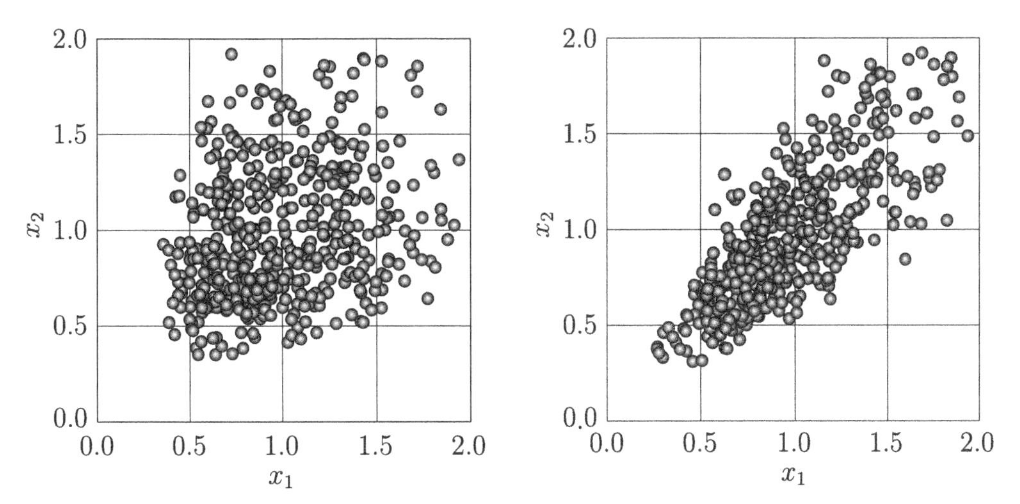

Figure 2.28 Monte Carlo samples of correlated lognormally distributed random variables (left: $\rho = 0.3$, right: $\rho = 0.8$).

2. Transform the sample into correlated standard Gaussian space by means of

$$\mathbf{v}^{(k)} = \mathbf{L}\mathbf{u}^{(k)} \tag{2.139}$$

Here, $\mathbf{L}$ is the Cholesky factor of the matrix of correlation coefficients ρ' as discussed in section 2.2.6.

3. Transform each variable separately into non-Gaussian space using

$$x_i^{(k)} = F_{X_i}^{-1}\left[\Phi(v_i^{(k)})\right] \tag{2.140}$$

Example 2.13 (Simulation of correlated lognormally distributed variables)
Consider two log-normally distributed random variables X_1, X_2 with identical means $\bar{X}_1 = \bar{X}_2 = \bar{X} = 1$ and identical standard deviations $\sigma_1 = \sigma_2 = \sigma = 0.4$. Assume further that the two variables are correlated with a coefficient of correlation ρ and that their joint probability density function is given by the Nataf-model. The transformation to Gaussian space and the computation of the adjusted coefficient of correlation is described in Example 2.5. The `octave` script for this example is given in Listing 2.6. The resulting samples are shown in Fig. 2.28 for $\rho = 0.3$ (left) and $\rho = 0.8$ (right).

Summary

This chapter presented the basic elements of probability theory as required for the modeling of random phenomena in terms of random variables and random vectors.

```
1  s=sqrt(log(1.16))
2  rho = 0.80
3  rhop=log(1+rho*s ^2)/log(1+s ^2)
4  mu=exp(-s ^2/2)
5  fid=fopen("simul08.txt", "w");
6  M=5000
7  R=[1,rhop;rhop,1];
8  L=chol(R)'
9  X1=zeros(M,1);
10 X2=zeros(M,1);
11 for i=1:M
12  u=randn(2,1);
13  v=L*u;
14  x1=mu*exp(v(1)*s);
15  x2=mu*exp(v(2)*s);
16  if (x1<2 && x2<2) fprintf(fid, "%g %g\n", x1,x2); end
17  X1(i)=x1;
18  X2(i)=x2;
19 end
20 fclose(fid);
21 plot(X1,X2,'o');
22 axis([0,2,0,2])
23 pause
```

Listing 2.6 Simulation of corrleated lognormally distributed random variables.

The definitions of expected values and probability distribution functions were introduced and related to each other. Various types of probability density function models were presented and compared to each other. The concept of conditional probability was presented and applied to conditional probability densities. The transformation rules for probability density functions of random variables related to each other by nonlinear mappings were derived. Multi-dimensional probability density models for random vectors were presented. The relationship between probabilistic models and real data was established in terms of estimation procedures with suitable properties. Confidence intervals and statistical test procedures enable the assessment of the model quality. A section on simulation techniques discussed various computational methods for generating random samples with prescribed statistical properties.

Chapter 3

Regression and response surfaces

ABSTRACT: The representation of complex physical or mathematical systems in terms of simple functional relations is an important issue in many areas. In an engineering context, it is desirable to describe the effect of uncertainty on the system's behavior in terms of the contributions due to individual random influences. This requires a statistically justified approximation model which is typically obtained by using regression techniques. This chapter introduces basic techniques for regression and for identification of important random variables alike. This is then complemented by a section on the appropriate choice of support points ("design of experiments").

3.1 Regression

It is frequently required to fit the parameters of a simple mathematical model describing a physical input-output relation to measured or computed sample data. The general concept of regression is to minimize the error in the sense of an expected value. Assume that the output quantity z is related to the n input quantities $x_1 \dots x_n$ that are assembled into a vector $\mathbf{x}$ by a functional relation $f(.)$ with the formal representation

$$z = f(\mathbf{p}, \mathbf{x}) \tag{3.1}$$

in which the function f depends on a parameter vector $\mathbf{p} = [p_1, p_2, \dots, p_\nu]^T$ whose values have yet to be determined. We assume that the available samples contain pairs $(\mathbf{x}^{(k)}, z^{(k)}), k = 1 \dots m$ of corresponding inputs and outputs. It is important to realize that these samples may contain random variability which is *not* described by the functional relation given in Eq. 3.1. Regression is carried out by minimizing the mean square difference (residual) S between the observed values $z^{(k)}$ and the predicted values $f(\mathbf{p}; \mathbf{x}^{(k)})$ by choosing an appropriate parameter vector $\mathbf{p}^*$.

$$S(\mathbf{p}) = \sum_{k=1}^{m} \left[z^{(k)} - f(\mathbf{x}^{(k)}, \mathbf{p}) \right]^2; \quad \mathbf{p}^* = \operatorname{argmin} S(\mathbf{p}) \tag{3.2}$$

For a linear regression in which the function f can be written as

$$f(\mathbf{p}, \mathbf{x}) = \sum_{i=1}^{\nu} p_i g_i(\mathbf{x}) \tag{3.3}$$

the problem can be solved directly. The necessary conditions for the existence of a local minimum are:

$$\frac{\partial S}{\partial p_j} = 0; \quad j = 1 \dots \nu \tag{3.4}$$

which, together with Eqs. 3.2 and 3.3, results in

$$\sum_{k=1}^{m} \left\{ g_j(\mathbf{x}^{(k)}) \left[z^{(k)} - \sum_{i=1}^{\nu} p_i g_i(\mathbf{x}^{(k)}) \right] \right\} = 0; \quad j = 1 \dots \nu \tag{3.5}$$

This is a system of linear equations for the parameter vector $\mathbf{p}$

$$\mathbf{Qp} = \mathbf{q} \tag{3.6}$$

in which the matrix $\mathbf{Q}$ and the vector $\mathbf{q}$ are defined by their elements

$$Q_{ij} = \sum_{k=1}^{m} g_i(\mathbf{x}^{(k)}) g_j(\mathbf{x}^{(k)}); \quad q_j = \sum_{k=1}^{m} z^{(k)} g_j(\mathbf{x}^{(k)}); \quad i,j = 1 \dots \nu \tag{3.7}$$

One may raise the question whether the model (here the linear regression model) is actually appropriate. One possibility for checking the adequacy of a model is the so-called coefficient of determination R^2. This is a statistical measure describing the correlation between the outputs as predicted by the regression model and the actual data. According to the definition of the correlation coefficient, this is

$$R^2 = \rho_{YZ}^2 = \left(\frac{\mathbf{E}[Y \cdot Z]}{\sigma_Y \sigma_Z} \right)^2; \; y = \sum_{i=1}^{\nu} p_i g_i(\mathbf{x}) \tag{3.8}$$

For a good model the value of R^2 should be close to 1. An alternate way of defining R^2 is given by

$$R^2 = 1 - \frac{S}{S_{tot}} = \frac{S_{tot} - S}{S_{tot}} \tag{3.9}$$

Here, S denotes the residual as defined in Eq. 3.2 and S_{tot} denotes the total sum of squares of the data:

$$S_{tot} = \sum_{k=1}^{m} [z^{(k)}]^2 \tag{3.10}$$

The residual S can also be expressed in terms of the coefficient of determination R^2 and the total sum of squares S_{tot}

$$S = S_{tot}(1 - R^2) \tag{3.11}$$

When considering a situation where the number of data samples is (almost) equal to the number of regression parameters, it becomes clear that the R^2 values tend to be over-optimistic in the sense that the model fits a small data sample including possibly present random disturbances. This will limit the predictive capabilities of the model since the random disturbances will then be different. Hence it is generally accepted to adjust the R^2 values such as to take into account the limited sample size (Wherry 1931)

$$R^2_{adj} = R^2 - \frac{\nu - 1}{m - \nu}(1 - R^2) \tag{3.12}$$

Here, m is the sample size and ν is the number of unknown parameters in the regression model. The adjusting correction term vanishes as the sample size m increases.

Example 3.1 (Fitting a polynomial function)
As an example, consider fitting a polynomial function

$$f(x) = \sum_{i=0}^{\nu-1} p_i x^i \tag{3.13}$$

to a given set of 6 data points as shown in Fig. 3.1. The value of ν varies from 2 (linear regression) to 5 (quartic regression). The sets of coefficients for these regressions are given in Table 3.1. A quintic regression fits perfectly through all data points since the number of sample points is equal to the number of regression parameters. However, the adjusted coefficient of determination R^2_{adj} cannot be computed. Comparing the values for R^2_{adj} in Table 3.1, it appears that a cubic model is best suited to represent the sample data.

Example 3.2 (Oscillator with random properties)
Consider a simple linear oscillator as shown in Fig. 3.2 with mass m, stiffness k and viscous damping c. It is subjected to a harmonic excitation with amplitude F_0 and circular frequency ω. The stationary response $x(t)$ is given by

$$x(t) = x_0 \sin(\omega t + \varphi) \tag{3.14}$$

with the phase angle φ and the amplitude

$$x_0 = \frac{F_0}{(k - m\omega^2)^2 + c^2\omega^2} \tag{3.15}$$

If we assume random properties for the system and load parameters, then the response amplitude obviously will be random as well. Assume that k, m, and c are log-normally distributed with mean values $\bar{k} = 2$, $\bar{m} = 1$, $\bar{c} = 0.01$ and ω is Gaussian with mean value $\bar{\omega} = 1$. The loading amplitude is assumed to be deterministic $F_0 = 1$. When carrying out a Monte-Carlo simulation with a coefficient of variation COV = 0.05 for the input variables, we obtain the regression results as shown in Fig. 3.3. It can be seen that the full quadratic model performs best, yet also the quadratic model without interaction terms and the linear model are quite satisfactory. Upon increasing the COV to 0.09, the changes are significant, as shown in Fig. 3.4. Now the randomness of the variables becomes significant in the sense that the regression model cannot represent the nonlinear relationship between input and output very well. Of course, here this is a consequence of the inadequacy of the regression model.

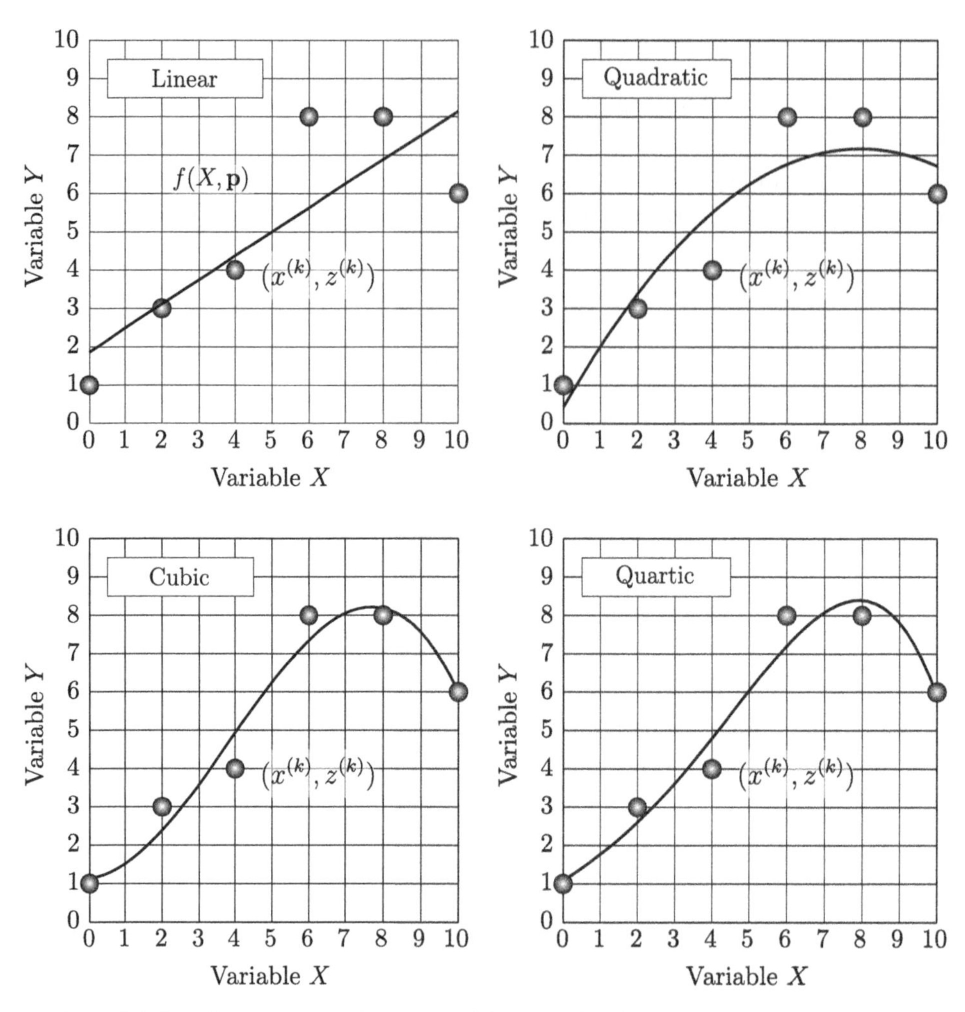

Figure 3.1 Set of data points with polynomial functions fitted by linear regression.

Table 3.1 Regression coefficients and R^2-values for polynomial regression.

ν	p_0	p_1	p_2	p_3	p_4	R^2	R^2_{adj}
2	1.857	0.629	–	–	–	0.69	0.61
3	0.429	1.700	−0.107	–	–	0.86	0.77
4	1.151	0.051	0.344	−0.030	–	0.96	0.89
5	1.079	0.646	0.024	0.022	−0.0026	0.96	0.80

3.2 Ranking of variables

Since the number of sample points should be significantly larger than the number of regression coefficients, it is quite important not to include irrelevant variables in

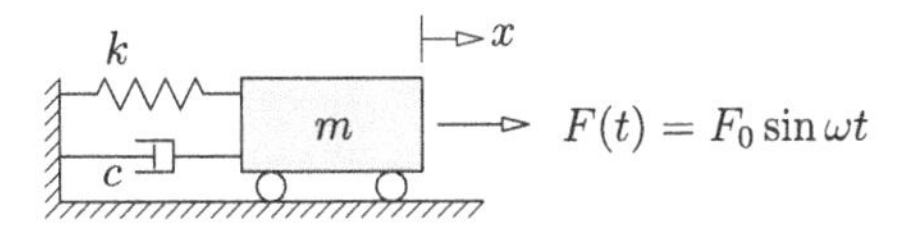

Figure 3.2 Simple oscillator.

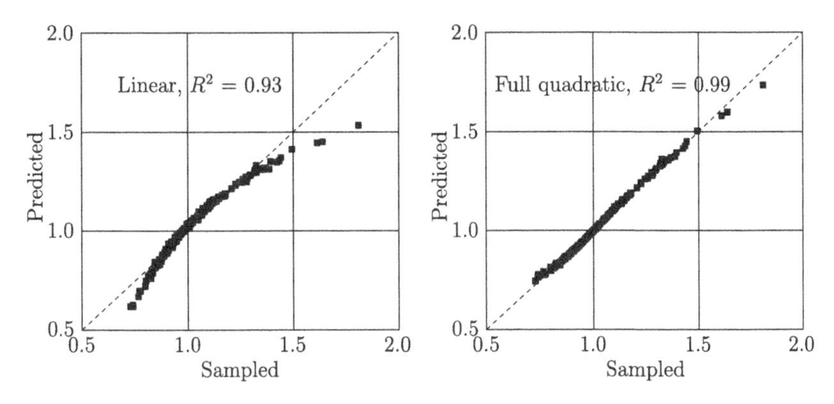

Figure 3.3 Regression for simple oscillator, COV of input = 0.05.

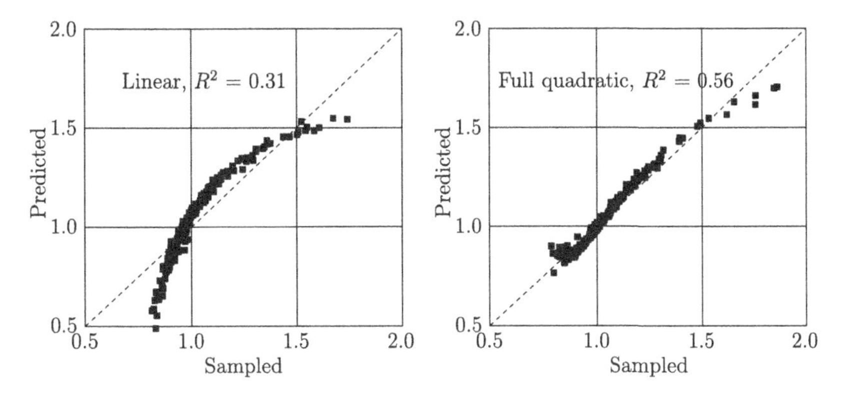

Figure 3.4 Regression for simple oscillator, COV of input = 0.09.

the regression model. A rather straightforward approach for ranking variables with respect to their importance in a regression model is to utilize the (adjusted) coefficient of determination. This is done by computing the coefficient of determination R^2 for the full model and compare it to the corresponding value R_e^2, which we obtain when when one variable x_e is removed from the regression one at a time. This removal will result in a drop

$$\Delta R_e^2 = R^2 - R_e^2 \tag{3.16}$$

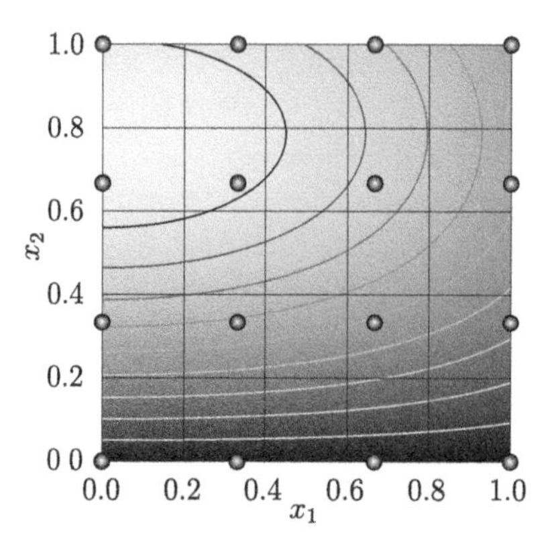

Figure 3.5 Simple sine function.

When we compare the values of ΔR_e^2 for all variables, we obtain an indicator of importance. Here it is helpful to normalize the values of ΔR_k^2 such that the maximum value is limited by 1. This defines a coefficient of importance

$$I_e = \frac{\Delta R_e^2}{R^2} = 1 - \frac{R_e^2}{R^2} \tag{3.17}$$

A similar comparison can be made on the basis of the adjusted values of the coefficient of determination. Also, the importance of individual coefficients p_e associated with one or more variables can be assessed in a similar manner.

Example 3.3 (Deterministic analytical function)
Consider the function $z = \cos x_1 \cdot \sin 2x_2$ on the square $[0, 1] \otimes [0, 1]$. Let the function be defined in 16 support points $(x_1^{(k)}, x_2^{(k)})$ as indicated in Fig. 3.5. The coefficients of correlation between the variables x_1, x_2 and the function z are $\rho_{x_1z} = -0.32$ and $\rho_{x_2z} = 0.83$. A quadratic model is fitted to these 16 data points

$$f = b_1 + b_2x_1 + b_3x_2 + b_4x_1^2 + b_5x_1x_2 + b_6x_2^2 \tag{3.18}$$

With this model, regression yields the coefficient vector

$$\mathbf{b} = [0.0016, 0.1928, 2.2224, -0.2679, -0.4266, -1.2531]^T \tag{3.19}$$

The coefficient of determination is $R^2 = 0.99$ and the adjusted value is $R_{adj}^2 = 0.988$. So this is indeed a very good representation of the data.

Carrying out a regression on a reduced model containing x_1 only, i.e.

$$f = b_1 + b_2x_1 + b_4x_1^2 \tag{3.20}$$

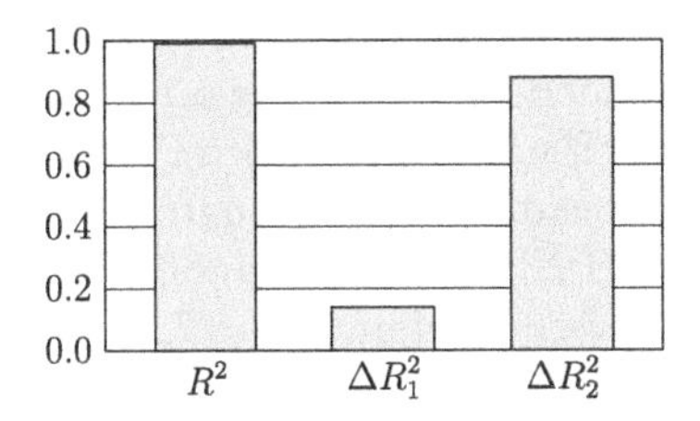

Figure 3.6 Importance measure based on coefficient of determination.

results in a coefficient of determination $R^2 = 0.11$ and a negative value of $R^2_{adj} = -0.03$. This indicates that the model is not useful. Performing a similar regression on a model containing x_2 only

$$f = b_1 + b_3 x_2 + b_6 x_2^2 \tag{3.21}$$

results in $R^2 = 0.85$ and $R^2_{adj} = 0.83$. This can still be considered acceptable. The big drop in the value of R^2 or R^2_{adj} when removing the variable x_2 from the regression is a good indicator of its relative importance compared to x_1 (cf. Fig. 3.3). If using a regression model containing linear terms in x_1 and x_2 only

$$f = b_1 + b_2 x_1 + b_3 x_2 \tag{3.22}$$

we obtain a coefficient of determination $R^2 = 0.78$ and $R^2_{adj} = 0.75$. So a linear model is not quite satisfactory.

As an alternative, we can apply Analysis of variance (ANOVA) techniques (Fisher 1921). The ANOVA procedure basically tests whether the mean values obtained from different sample sets are identical. In this context, it is applied in such a way that the predictions obtained from two different models, e.g. by the full model (3.3) and by a reduced model eliminating one coefficient p_e (3.23), are compared

$$f(\mathbf{p}; \mathbf{x}) = \sum_{i=1, i \neq e}^{\nu} p_i g_i(\mathbf{x}) \tag{3.23}$$

This will lead to a residual sum of squares S_e^2, which will be larger than the minimum obtained from Eq. 3.2. The quantity

$$F_e = \frac{1}{m - \nu} \frac{S_e^2 - S^2}{S^2} \tag{3.24}$$

is a random variable whose distribution follows an F distribution. The hypothesis to be tested in this context is whether $F \neq 0$ with statistical significance. In order to perform the test, F is compared with the critical value, the F-statistic $F_c = F_{\alpha,1,m-\nu}$, so that if $F_e > F_{\alpha,1,m-\nu}$, p_e we may conclude that it is non-zero with confidence $1 - \alpha$. Also, the F-values can be used to identify important coefficients (those which have non-zero values with a larger confidence). Alternatively, the magnitude of the confidence in the hypothesis $p_e \neq 0$ can be obtained by computing α from $F = F_{\alpha,1,m-\nu}$. For the previous example, we have $m - \nu = 10$. If we set $\alpha = 0.05$ we have $F_c = F_{0.05,1,10} = 4.96$. For the full quadratic model we get a residual sum of squares $S^2 = 0.02$, for the model without the variable x_1 we have $S_1^2 = 0.28$ and for the model without the variable x_2 we get $S_2^2 = 1.66$. The test values are $F_1 = 1.14$ and $F_2 = 7.33$. This way, at the given level of significance we have to conclude that the variable x_1 should be omitted ($F_1 < F_c$) and the variable x_2 should be included ($F_2 > F_c$).

When we rewrite Eq. 3.24 in terms of the relations (3.9), we can see that

$$F_e = \frac{1}{m-\nu} \frac{S_{tot}(1-R_e^2) - S_{tot}(1-R^2)}{S_{tot}(1-R^2)} = \frac{1}{m-\nu} \frac{R^2 - R_e^2}{1-R^2} = \frac{1}{m-\nu} \frac{\Delta R_e^2}{1-R^2} \quad (3.25)$$

From this, it can be noted that the F-values, when normalized suitably, match the corresponding results obtained from the coefficient of determination as given in Eq. 3.16.

Example 3.4 (Oscillator with random properties)
Consider the linear oscillator as discussed in Example 3.2. Based on the full quadratic regression models, the influence of the random variables on the structural response (for different values of their coefficients of variation) can be found as indicated in Table 3.2.

This example shows that using the adjusted value of the coefficient of determination may lead to slightly negative values of the coefficient of importance. This is caused by the way R^2_{adj} is constructed and should be interpreted as an indicator that the variable in question may safely be omitted from the analysis. In this case, the damping constant c is irrelevant. This is due to the fact that the natural circular frequency, as computed from the mean values of m and k, is located at $\omega_0 = \sqrt{\frac{k}{m}} = 1.414$ rad/s, which is relatively far away from the mean value of the excitation frequency at 1 rad/s. So resonance effects (in which case the value of the damping constant would be very influential) are unlikely to occur.

Table 3.2 Ranking of input variables for random oscillator based on R^2_{adj}.

Variable	I_e	
	COV = 0.05	*COV = 0.09*
m	0.1344	0.2363
c	0.0000	−0.0003
k	0.4358	0.3886
ω	0.4825	0.6116

3.3 Response surface models

3.3.1 *Basic formulation*

Response surface models are more or less simple mathematical models which are designed to describe the possible experimental outcome (e.g., the structural response in terms of displacements, stresses, etc.) of a more or less complex structural system as a function of variable factors (e.g., loads or system conditions), which can be controlled by an experimenter. Obviously, the chosen response surface model should give the best possible fit to any collected data. In general, we can distinguish two different types of response surface models:

- regression models (e.g., polynomials of varying degree or non-linear functions such as exponentials)
- interpolation models (e.g., polyhedra, radial basis functions).

Let us denote the response of any structural system to n experimental factors or input variables x_i; $i = 1 \ldots n$, which are assembled into a vector $\mathbf{x} = (x_1 \ldots x_n)^T$, by $z(\mathbf{x})$. In most applications it is quite likely that the *exact* response function will not be known. Therefore, it has to be replaced by a sufficiently versatile function $\eta(\cdot)$ which will express the relation between the response z and the input variables $\mathbf{x}$ satisfactorily. If we take into account a zero-mean random error term ε, then the response can be written over the region of experimentation as

$$z = \eta(p_1 \ldots p_\upsilon;\ x_1 \ldots x_n) + \varepsilon \tag{3.26}$$

in which p_j; $j = 1 \ldots p$ are the parameters of the approximating function $\eta(\cdot)$. We now apply the expectation operator, i.e.,

$$\eta = \mathbf{E}[z] \tag{3.27}$$

Then the surface represented by

$$\eta(p_1 \ldots p_\upsilon;\ x_1 \ldots x_n) = \eta(\mathbf{p};\ \mathbf{x}) \tag{3.28}$$

is called a *response surface* (cf. Fig. 3.7). The vector of parameters $\mathbf{p} = (p_1 \ldots p_\upsilon)^T$ has to be estimated from the experimental data in such a way that Eq. (3.27) is fulfilled. In the following, we will investigate the most common response surface models and methods to estimate their respective parameters.

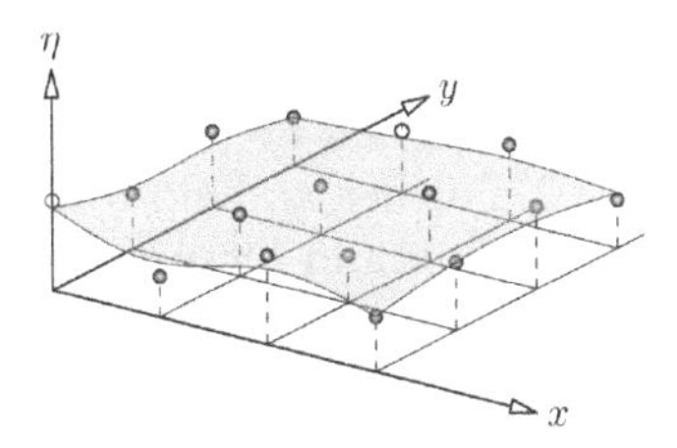

Figure 3.7 Response surface and data points.

3.3.2 *Linear models and regression*

Let us assume that an appropriate response surface model $\eta(\cdot)$ has been chosen to represent the experimental data. Then, for estimating the values of the parameters $\boldsymbol{\theta}$ in the model, the method of maximum likelihood can be utilized. Under the assumptions of a Gaussian distribution of the random error terms ε, the method of maximum likelihood can be replaced by the more common method of least squares (Box and Draper 1987). In the latter case the parameters $\mathbf{p}$ are determined in such a way that the sum of squares of the differences between the value of the response surface $\eta(\mathbf{p};\mathbf{x}^{(k)})$ and the measured response $z^{(k)}$ at the m points of experiment

$$\mathbf{x}^{(k)} = [x_1^{(k)}, \ldots, x_n^{(k)}]^T, \quad k = 1 \ldots m \tag{3.29}$$

becomes as small as possible. In other words, the sum of squares function

$$S^2 = \sum_{k=1}^{m} \left[z^{(k)} - \eta(\mathbf{p};\mathbf{x}^{(k)}) \right]^2 \tag{3.30}$$

has to be minimized (cf. Eq. 3.2). This corresponds to a minimization of the variance of the random error terms ε. The minimizing choice of $\mathbf{p}$ is called a *least-squares estimate* and is denoted by $\mathbf{p}^*$.

The above regression problem becomes quite simple if the response surface model is linear in its parameters $\mathbf{p}$ as defined in Eq. 3.3. Then, the least squares estimate $\mathbf{p}^*$ of the parameter vector is determined by the linear system of equations (3.6). This estimator is unbiased, i.e.,

$$\mathbf{E}[\mathbf{p}^*] = \mathbf{p} \tag{3.31}$$

Assuming that the covariance of the observations $z^{(k)}$ is given by

$$\mathbf{C_{zz}} = [\mathbf{z}] = \mathbf{E}[(\mathbf{z} - \mathbf{E}[\mathbf{z}])(\mathbf{z} - \mathbf{E}[\mathbf{z}])^T] = \sigma^2\mathbf{I} \tag{3.32}$$

then the covariance matrix of the parameters is

$$\mathbf{C_{pp}} = \mathbf{E}[(\mathbf{p}^* - \mathbf{p})(\mathbf{p}^* - \mathbf{p})^T] = \sigma^2(\mathbf{Q}^T\mathbf{Q})^{-1} \tag{3.33}$$

in which $\mathbf{Q}$ has been defined in Eq. 3.7. Since a response surface is only an approximation of the functional relationship between the structural response and the basic variables, it should be evident that, in general, there is always some *lack of fit* present. Therefore, a crucial point when utilizing response surfaces is to check whether the achieved fit of the response surface model to the experimental data suffices or if the response surface model has to be replaced by a more appropriate one. This follows the discussion given in section 3.2. Further, and more advanced, measures or checking procedures can be found, e.g., in Box and Draper 1987; Khuri and Cornell 1996; Myers and Montgomery 2002; Böhm and Brückner-Foit 1992. A detailed discussion is given by Bucher and Macke 2005; Bucher and Most 2008.

It should be pointed out that the estimated parameters $\mathbf{p}^*$ are least square estimates, i.e., there is a certain likelihood that the true parameter $\mathbf{p}$ has a different value than the estimated one. Therefore, it is sometimes quite advisable to determine confidence intervals for the parameters (cf. section 2.3.2). Joint confidence regions for several regression coefficients are given, e.g., in Myers and Montgomery 2002. Consequently, when utilizing response surfaces in reliability assessment, we should be aware that all predictions—may it be the structural response for a certain design or a reliability measure—show respective prediction uncertainty as reflected in confidence intervals.

3.3.3 *First- and second-order polynomials*

As already mentioned above, response surfaces are designed in a way that a complex functional relation between the structural response and the basic variables is described by an appropriate, but—preferably—as simple as possible mathematical model. In the context of response surfaces, the term "simple" means that the model should be continuous in the basic variables and have a small number of terms whose coefficients can be easily estimated. Polynomial models of low-order fulfill such demands. Therefore, in the area of stochastic structural analysis (e.g. for reliability assessment) the most common response surface models are first- and second-order polynomials (cf. Rackwitz 1982, Bucher and Bourgund 1990; Kim and Na 1997; Zheng and Das 2000).

The general form of a first-order model of a response surface η which is linear in its n basic variables x_i is

$$\eta = p_0 + \sum_{i=1}^{n} p_i x_i \tag{3.34}$$

with p_i; $i = 0, 1 \ldots n$ as the unknown parameters to be estimated from the experimental data. This model has a total of $\nu = n + 1$ parameters. The parameter p_0 is the value of the response surface at the origin or the center of the experimental design, whereas the coefficients p_i can be interpreted as the gradients of the response surface in the direction of the respective basic variables x_i. As can be seen in Eq. (3.34), the first-order model is not able to represent even the simplest interaction between the input variables.

If it becomes evident that the experimental data can not be represented by a model whose basic variables do not have mutually independent effects, then the first-order model can be enriched with (simple) interaction terms, so that

$$\eta = \theta_0 + \sum_{i=1}^{n} p_i x_i + \sum_{i=1}^{n-1} \sum_{j=i+1}^{n} p_{ij} x_i x_j \tag{3.35}$$

The total number of parameters to be estimated is given by $\nu = 1 + n(n+1)/2$. In the response surface model of Eq. (3.35) there is some curvature present but that only results from the twisting of the planes of the respective input variables. If a substantial curvature is required as well, then the above model can be further enhanced by n quadratic terms to a complete second-order model in the form of

$$\eta = \theta_0 + \sum_{i=1}^{n} p_i x_i + \sum_{i=1}^{n} \sum_{j=i}^{n} p_{ij} x_i x_j \tag{3.36}$$

The total number of parameters to be estimated is $v = 1 + n + n(n+1)/2$.

In most common cases either the first-order or the complete second-order model are utilized as response surface functions.

3.3.4 *Weighted interpolation*

Besides the above mentioned regression models, interpolation models are also available for describing response surfaces.

A useful Shepard-type (Shepard 1968) interpolation strategy (particularly for reliability problems with convex limit state functions) has been described by Bucher, Hintze, and Roos 2000. It appears to be helpful to carry out the interpolation in non-dimensional space (e.g., standard Gaussian space $\mathbf{u}$). Shepard interpolation requires a distance measure d which is most suitably defined in terms of the Euclidian distance between two vectors $\mathbf{u}_\ell$ and $\mathbf{u}_k$:

$$d(\mathbf{u}_\ell, \mathbf{u}_k) = \|\mathbf{u}_\ell - \mathbf{u}_k\| \tag{3.37}$$

For a given vector $\mathbf{U}$, the function $g(\mathbf{U})$ is approximated by

$$\eta(U) = \frac{\sum_{i=1}^{m} w_i g(\mathbf{u}_i)}{\sum_{i=1}^{m} w_i} \tag{3.38}$$

in which the weights w_i are computed from

$$w_i = [d(\mathbf{U}, \mathbf{u}_i) + \epsilon]^{-p} \tag{3.39}$$

Here, ϵ is a small positive number regularizing this expression and p is a suitably chosen positive number (e.g. $p = 2$).

All these Shepard type models allow, by including additional points, an adaptive refinement of the approximating response surface.

Example 3.5 (Shepard interpolation of 2D function over irregular set of support points)
Consider a function $g : \mathbb{R}^2 \to \mathbb{R}$ defined in 5 points as given in Table 3.3. We want to construct a Shepard interpolation function over the square $[0,1] \otimes [0,1]$. The location of the 5 support points as well as the contour lines of the interpolated function η are shown in Fig. 3.8.

Table 3.3 Data for Shepard interpolation in $\mathbb{R}^2$.

u_1	u_2	g
0.21307	0.87950	0.164813
0.84100	0.71685	0.432167
0.63773	0.24935	0.039652
0.54003	0.37962	0.077823
0.73806	0.64272	0.304883

3.3.5 *Moving least squares regression*

An arbitrary function u is interpolated at a point $\mathbf{x}_k$ by a polynomial as

$$u_k = u(\mathbf{x}_k) = [1\ x_k\ y_k \dots x_k^2\ x_k y_k\ y_k^2\ \dots] \begin{bmatrix} a_{1k} \\ \vdots \\ a_{\nu k} \end{bmatrix} = \mathbf{p}^T(\mathbf{x}_k)\mathbf{a}_k = \mathbf{p}_k^T \mathbf{a}_k \tag{3.40}$$

where $\mathbf{p}(\mathbf{x}_k)$ is the vector of base monomials evaluated at the point $\mathbf{x}_k$ and $\mathbf{a}_k$ contains the coefficients of the polynomial. These coefficients are usually assumed to be constant in the interpolation domain and can be determined directly if the number of supporting points m used for the interpolation is equal to the number of coefficients ν.

Within the "Moving Least Squares" (MLS) interpolation method (Lancaster and Salkauskas 1981) the number of supporting points m exceeds the number of coefficients n, which leads to an overdetermined system of equations. This can be formulated as an optimization problem to be solved by using a least squares approach

$$\mathbf{P}\mathbf{u} = \mathbf{P}\mathbf{P}^T\mathbf{a}(\mathbf{x}) \tag{3.41}$$

with changing ("moving") coefficients $\mathbf{a}(\mathbf{x})$. Here $\mathbf{u}$ denotes the assembly of all function values u_k into a vector, and $\mathbf{P}$ denotes the assembly of all vectors $\mathbf{p}_k$ into a matrix. If an interpolation technique should by used in a numerical method the compact support of the interpolation is essential for an efficient implementation. This was realized for the MLS-interpolation by introducing a distance depending weighting function $w = w(s)$, where s is the standardized distance between the interpolation point and the considered supporting point

$$s_i = \frac{\|\mathbf{x} - \mathbf{x}_i\|}{D} \tag{3.42}$$

and D is the influence radius, which is defined as a numerical parameter. All types of functions can be used as weighting functions $w(s)$ which have their maximum in $s = 0$ and vanish outside of the influence domain specified by $s = 1$.

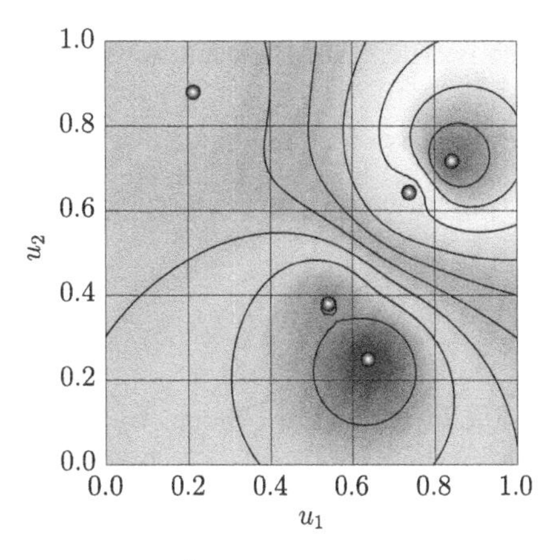

Figure 3.8 Shepard interpolation in $\mathbb{R}^2$.

The final approximation scheme then becomes

$$\begin{aligned} \eta(\mathbf{x}) &= \mathbf{\Phi}^{MLS}(\mathbf{x})\mathbf{u} \\ \mathbf{\Phi}^{MLS}(\mathbf{x}) &= \mathbf{p}^T(\mathbf{x})\mathbf{A}(\mathbf{x})^{-1}\mathbf{B}(\mathbf{x}) \end{aligned} \tag{3.43}$$

with

$$\begin{aligned} \mathbf{A}(\mathbf{x}) &= \mathbf{P}\mathbf{W}(\mathbf{x})\mathbf{P}^T \\ \mathbf{B}(\mathbf{x}) &= \mathbf{P}\mathbf{W}(\mathbf{x}) \end{aligned} \tag{3.44}$$

In Most and Bucher 2005 a new weighting function was presented which enables the fulfillment of the MLS interpolation condition with very high accuracy without any additional numerical effort

$$\Phi_i^{MLS}(\mathbf{x}_j) \approx \delta_{ij} \tag{3.45}$$

This can only be reached if Eq. (3.46) holds

$$w_i(\mathbf{x}_j) \approx \delta_{ij} \tag{3.46}$$

The weighting function value of a node i at an interpolation point $\mathbf{x}$ is introduced by the following regularized formulation

$$w_R(s_i) = \frac{\tilde{w}_R(s_i)}{\sum_{j=1}^m \tilde{w}_R(s_j)} \tag{3.47}$$

with

$$\tilde{w}_R(s) = (s^2 + \epsilon)^{-2}; \quad \epsilon \ll 1 \tag{3.48}$$

The regularization parameter ϵ has to be chosen small enough to fulfill Eq. (3.46) with high accuracy but large enough to obtain a regular, differentiable function at $s = 0$ within the machine precision. This approach allows for a very good representation of given support points which is crucial for the accuracy of reliability estimates.

3.3.6 *Radial basis functions*

Radial basis function interpolation constructs interpolating functions of the form

$$\eta(\mathbf{x}) = \sum_{j=1}^{m} p_j \varphi(\|\mathbf{x} - \mathbf{x}^{(j)}\|) \tag{3.49}$$

in which $\|.\|$ denotes the Euclidian norm of the vector argument and $\varphi(.)$ are usually taken to be simple functions of their arguments, e.g. linear: $\varphi(r) = cr$, "thin plate spline": $\varphi(r) = cr^2 \log r$, or Gaussian: $\varphi(r) = \exp(-cr^2)$ (Buhmann 2004). The vectors $\mathbf{x}_j$ denote the support points in which the function is given, and the coefficients a_j have to be adjusted such as to fulfil the interpolation condition. Since this is an interpolation method, there are exactly as many coefficients as there are data points. The coefficients are then to be determined from a system of linear equations

$$z^{(k)} = z(\mathbf{x}^{(k)}) = \sum_{j=1}^{m} p_j \varphi(||\mathbf{x}^{(k)} - \mathbf{x}^{(j)}||); \; k = 1 \ldots m \tag{3.50}$$

A major advantage of the radial basis function interpolation is that the matrix in Eq. 3.50 is always non-singular. This implies that there is always a unique solution (Buhmann 2004). Due to the radial symmetry of the basis functions it is advantageous to introduce a scaling of the independent variables such that they all are in the same range of values, e.g. in the interval $[0, 1]$. A detailed discussion of this methodology is given by Buhmann 2004. Shepard interpolation (cf. section 3.3.4) is a special, very simple case of radial basis function interpolation.

Another point of relevance for some applications is the fact that radial basis functions are unable to represent simple but important functions such as constants or linear functions exactly. One way of including these types of functions into the analysis is to "stack" the radial basis function interpolation on top of a linear regression with $n+1$ coefficients p'_k; $k=0 \ldots n$ so that

$$z^{(k)} = p'_0 + \sum_{k=1}^{n} p'_k x_k^{(j)} + \sum_{j=1}^{m} p_j \varphi(\|\mathbf{x} - \mathbf{x}^{(j)}\|) \tag{3.51}$$

in which the regression model is determined beforehand, and then the coefficients p_j are computed from the difference between the actual data and linear the regression model.

Example 3.6 [Radial basis function interpolation]
Consider a function defined in 6 data points as shown in Table 3.4. We use the thin plate spline function $\varphi(r) = cr^2 \log r$ as basis function. For the linear regression we need to solve the system of equations (cf. Eq. 3.6)

$$\begin{bmatrix} 6.000 & 3.000 & 3.000 \\ 3.000 & 2.265 & 1.625 \\ 3.000 & 1.625 & 2.265 \end{bmatrix} \cdot \begin{bmatrix} p'_0 \\ p'_1 \\ p'_2 \end{bmatrix} = \begin{bmatrix} 9.000 \\ 5.250 \\ 6.050 \end{bmatrix} \tag{3.52}$$

Table 3.4 Sample data for radial basis function.

j	$x_1^{(j)}$	$x_2^{(j)}$	$z^{(j)}$
1	0.1	0.1	1
2	0.9	0.1	1
3	0.9	0.9	2
4	0.1	0.9	2
5	0.25	0.25	0
6	0.75	0.75	3

with the solution

$$\begin{bmatrix} p'_0 \\ p'_1 \\ p'_2 \end{bmatrix} = \begin{bmatrix} 0.208 \\ 0.667 \\ 1.917 \end{bmatrix} \tag{3.53}$$

The system of equations to be solved for the radial basis function part is then

$$\begin{bmatrix} 0.000 & -0.143 & 0.158 & -0.143 & -0.070 & -0.071 \\ -0.143 & 0.000 & -0.143 & 0.158 & -0.180 & -0.180 \\ 0.158 & -0.143 & 0.000 & -0.143 & -0.071 & -0.070 \\ -0.143 & 0.158 & -0.143 & 0.000 & -0.180 & -0.180 \\ -0.070 & -0.180 & -0.071 & -0.180 & 0.000 & -0.173 \\ -0.071 & -0.180 & -0.070 & -0.180 & -0.173 & 0.000 \end{bmatrix} \cdot \begin{bmatrix} p_1 \\ p_2 \\ p_3 \\ p_4 \\ p_5 \\ p_6 \end{bmatrix} = \begin{bmatrix} 0.534 \\ -0.000 \\ -0.534 \\ 0.000 \\ -0.854 \\ 0.854 \end{bmatrix} \tag{3.54}$$

with the solution

$$\begin{bmatrix} p_1 \\ p_2 \\ p_3 \\ p_4 \\ p_5 \\ p_6 \end{bmatrix} = \begin{bmatrix} -3.421 \\ 0.000 \\ 3.421 \\ -0.000 \\ -4.901 \\ 4.901 \end{bmatrix} \tag{3.55}$$

Applying these coefficients for interpolation of the function over the square $[0,1] \otimes [0,1]$, results in a smooth surface as indicated in Fig. 3.9.

The `octave` script solving the example is shown in Listing 3.1.

Exercise 3.1 (Interpolation of trigonometric function)
Perform a radial basis function interpolation $\eta(x_1,x_2)$ of the trigonometric function $z(x_1,x_2) = \sin \pi x_1 \cdot \sin 2\pi x_2$ over the square $[0,1] \otimes [0,1]$ using the support points as shown in Fig. 3.10. Compute the function values of η at the point $\mathbf{x}_r = [0.1, 0.1]^T$.

Solution: The coefficients for the linear regression part are

$$\begin{bmatrix} p'_0 \\ p'_1 \\ p'_2 \end{bmatrix} = \begin{bmatrix} 0.283 \\ 0.000 \\ -0.566 \end{bmatrix} \tag{3.56}$$

and the coefficients for the radial basis functions, numbered according to the support points fiven in Fig. 3.10, are

$$\begin{bmatrix} p_1 \\ p_2 \\ p_3 \\ p_4 \\ p_5 \\ p_6 \\ p_7 \\ p_8 \\ p_9 \end{bmatrix} = \begin{bmatrix} -0.391 \\ -0.391 \\ 2.822 \\ 2.822 \\ 0.000 \\ -2.822 \\ -2.822 \\ 0.391 \\ 0.391 \end{bmatrix} \tag{3.57}$$

```
 1  xyz=[0.1,0.1,1;
 2  .9,0.1,1;
 3  .9,.9,2;
 4  .1,.9,2;
 5  .25,.25,0;
 6  .75,.75,3];
 7  xy = xyz(:,1:2);
 8  z = xyz(:,3);
 9  p=[ones(6,1),xy];
10  mat1=p'*p;
11  rhs1=p'*z;
12  coeff1=mat1\rhs1
13  z1=p*coeff1;
14  z=z-z1;
15  mat=zeros(6,6);
16  for i=1:6
17    for k=1:6
18      x=xy(i,:)';
19      y=xy(k,:)';
20        r = sqrt((x(1)-y(1))^2+(x(2)-y(2))^2);
21        l = 0;
22        if (r>0) l = r^2*log(r); end
23      mat(i,k) = l;
24      end
25    end
26  coeff=mat\z
```

Listing 3.1 Computation of coefficients for radial basis function interpolation.

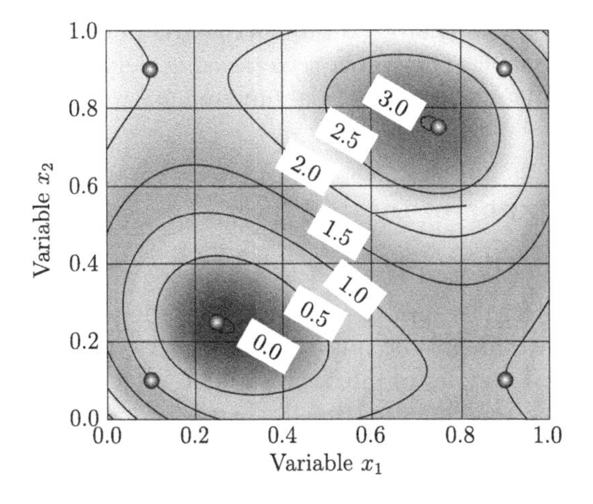

Figure 3.9 Radial basis function interpolation.

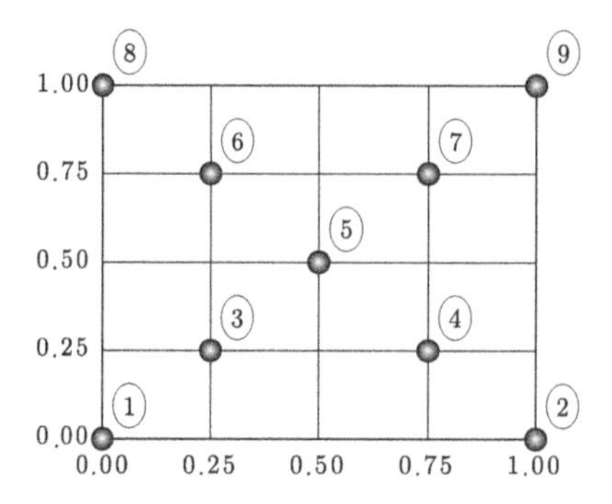

Figure 3.10 Support points for Exercise 3.1.

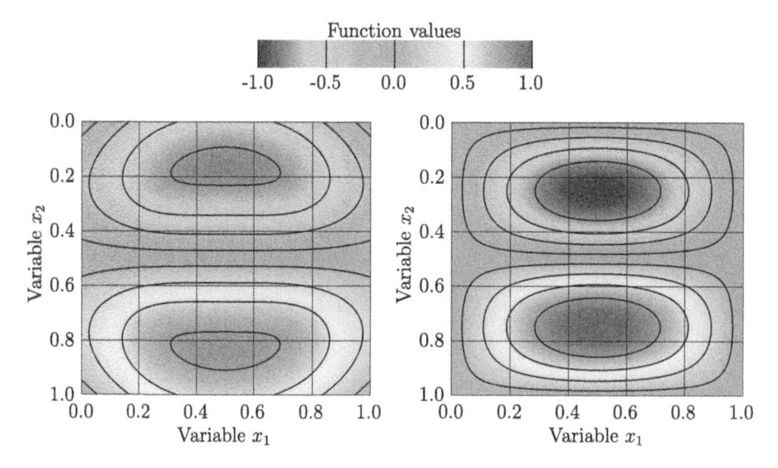

Figure 3.11 RBF interpolation for Exercise 3.1, interpolation (left) vs. original function (right).

and the interpolated function value at the requested point is $\eta(\mathbf{x}_r) = 0.398$. Note that the true value is $z(\mathbf{x}_r) = 0.1816$. Closer inspection reveals that in some locations the difference is even considerably larger although the overall behavior of the interpolated function is quite reasonable. This can be seen from Fig. 3.11 which compares the radial basis function interpolation to the original function.

3.4 Design of experiments

3.4.1 *Transformations*

Depending on the selected response surface model, support points $\mathbf{x}^{(k)}, k = 1, \ldots, m$ have to be chosen to estimate the unkown parameters of the response surface in a sufficient way. A set of samples of the basic variables is generated for this purpose. In general, this is done by applying predefined schemes, so called *designs of experiments*. The schemes shown in the following are saturated designs for first- and second-order polynomials, as well as full factorial and central composite designs. As is quite well

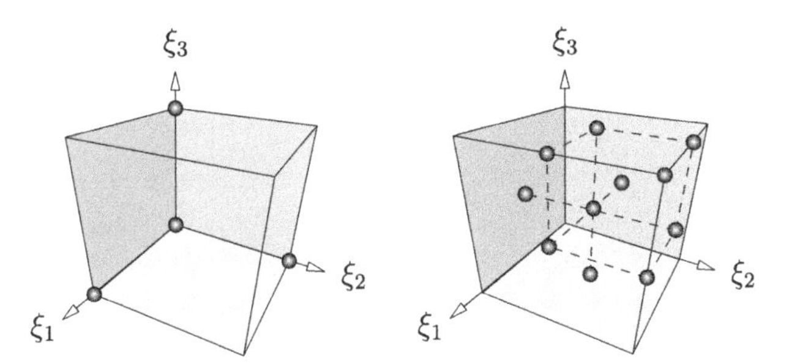

Figure 3.12 Saturated linear experimental scheme (left) and quadratic design scheme for $n = 3$ (right).

known from experiments regarding physical phenomena, it is most helpful to set up the experimental scheme in a space of dimensionless variables. The schemes, as they are described in the following, perform experimental designs in a space of dimension n, where n is equal to the number of relevant basic variables.

The selected design of experiments provides us with a grid of points defined by the dimensionless vectors $\boldsymbol{\xi}^{(k)} = [\xi_1^{(k)}, \xi_2^{(k)}, \ldots, \xi_n^{(k)}]^T$. This grid has to be centered around a vector $\mathbf{c} = [c_1, c_2, \ldots, c_n]^T$. In the absence of further knowledge, this center point can be chosen equal to the vector of mean values $\boldsymbol{\mu} = [\mu_1, \mu_2, \ldots, \mu_n]^T$ of the basic random variables $X_i, i = 1, \ldots, n$. As further knowledge about the actual problem becomes available other choices of the center point are, in general, more appropriate. The distances from the center are controlled by the scaling vector $\mathbf{s} = [s_1, s_2, \ldots, s_n]^T$. In many cases it is useful to choose the elements s_i of this scaling vector equal to the standard deviations σ_i of the random variables X_i. So, in general, a support point $\mathbf{x}^{(k)}, k = 1, \ldots, m$ is defined as

$$\mathbf{x}^{(k)} = \begin{bmatrix} x_1^{(k)} \\ x_2^{(k)} \\ \vdots \\ x_n^{(k)} \end{bmatrix} = \begin{bmatrix} c_1 + \xi_1^{(k)} s_1 \\ c_2 + \xi_2^{(k)} s_2 \\ \vdots \\ c_n + \xi_n^{(k)} s_n \end{bmatrix} \tag{3.58}$$

The number m of generated support points depends on the selected method. For the specific needs of reliability analysis (cf. section 6.3) it turns out to be very important to choose the scaling factors $\mathbf{s} = [s_1, s_2, \ldots, s_n]^T$ in such a way that the support points lie on the limit state surface, i.e. $g(\mathbf{x}^{(k)}) = 0$.

3.4.2 *Saturated designs*

Saturated designs provide a number of support points that just suffice to represent a certain class of response functions exactly. Hence for a linear saturated design, a linear function will be uniquely defined. Obviously, $m = n + 1$ samples are required for this purpose (cf. Fig. 3.12), and there is some arbitrariness in the design scheme which can usually be resolved only by introducing additional knowledge about the system behavior.

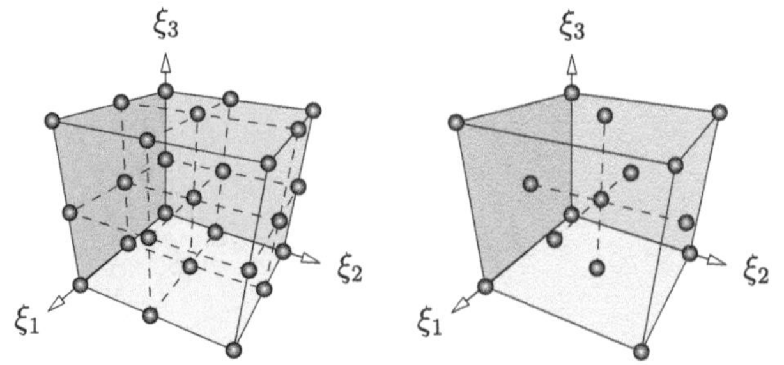

Figure 3.13 Full factorial design scheme for $q = 3$ and $n = 3$ (left), central composite design scheme for $n = 3$ (right).

A saturated quadratic design (Fig. 3.12) generates $m = n(n+1)/2 + n + 1$ support points $\mathbf{x}^{(k)}$. Any change of sign in the pairwise combinations would also lead to a saturated design, so that the final choice is somewhat arbitrary and should be based on additional problem-specific information.

3.4.3 *Redundant designs*

Redundant experimental design methods provide more support points than required to define the response surface, and thus enable error checking procedures as outlined in the preceding section. Typically, regression is used to determine the coefficients of the basis function. Here, linear and quadratic functions are also utilized.

The *full factorial* method (Fig. 3.13) generates q sample values for each coordinate, thus produces a total of $m = q^n$ support points $\mathbf{x}^{(k)}$ $(k = 1, 2, \ldots, m)$. Note that even for moderate values of q and n this may become prohibitively expensive. Therefore, subsets are frequently chosen, which leads to *fractional factorial* designs.

The *central composite* design method (Fig. 3.13) superimposes a full factorial design with $q = 2$ and a collection of all center points on the faces of an n-dimensional hypercube. Thus it generates $m = (2^n + 2n)$ support points $\mathbf{x}^{(k)}$.

D-optimal designs attempt to maximize the information content if only a small subset of the, otherwise preferable, full factorial design can be utilized, e.g., due to restrictions on computer capacity. Given a set of candidate factors $\boldsymbol{\xi}^{(k)}$ a subset of size m' is chosen in order to maximize the determinant of the Fisher information matrix $\mathbf{I}_F$

$$D = \det \mathbf{I}_F = \det (\mathbf{G}^T \mathbf{G}) \tag{3.59}$$

In this equation, $\mathbf{G}$ denotes a matrix containing values of the basis functions $g_i(\mathbf{x})$ for the response surface evaluated at the selected support points. Typically, the number m' is chosen to be 1.5-times the corresponding number of a saturated design.

Example 3.7 [*D*-optimal design in 2D]

Assume that we want to set up an experimental design for a linear response surface involving two variables x_1 and x_2 over the square $[0, 1] \otimes [0, 1]$. We have the basis functions 1, x_1, and x_2. Hence the saturated design needs $m = 3$ support points. A D-optimal design with $m' = 5$ is defined in terms of

$$\mathbf{G} = \begin{bmatrix} 1 & 1 & 1 & 1 & 1 \\ x_1^{(1)} & x_1^{(2)} & x_1^{(3)} & x_1^{(4)} & x_1^{(5)} \\ x_2^{(1)} & x_2^{(2)} & x_2^{(3)} & x_2^{(4)} & x_2^{(5)} \end{bmatrix}^T = \begin{bmatrix} 1 & 1 & 1 & 1 & 1 \\ x_{11} & x_{12} & x_{13} & x_{14} & x_{15} \\ x_{21} & x_{22} & x_{23} & x_{24} & x_{25} \end{bmatrix}^T \qquad (3.60)$$

obtained from maximizing

$$\begin{aligned} D &= \det \mathbf{G}^T\mathbf{G} = \\ &= \left(x_{15}^2 + x_{14}^2 + x_{13}^2 + x_{12}^2 + x_{11}^2\right)\left(x_{25}^2 + x_{24}^2 + x_{23}^2 + x_{22}^2 + x_{21}^2\right) \\ &\quad - (x_{15}\,x_{25} + x_{14}\,x_{24} + x_{13}\,x_{23} + x_{12}\,x_{22} + x_{11}\,x_{21})^2 \end{aligned} \qquad (3.61)$$

Choosing 4 support points at the corners of the square, we are left with the reduced problem

$$D = 4\,x_{25}^2 - 4\,x_{25} + 4\,x_{15}^2 - 4\,x_{15} + 7 \rightarrow \text{Max.!} \qquad (3.62)$$

from which we get the conditions

$$\begin{aligned} \frac{\partial D}{\partial x_1} &= 8\,x_{15} - 4 = 0 \\ \frac{\partial D}{\partial x_2} &= 8\,x_{25} - 4 = 0 \end{aligned} \qquad (3.63)$$

This system of equations has the solution $x_1^{(5)} = x_2^{(5)} = 0.5$.

Summary

This chapter presented techniques for deriving models based on observed or computed data. Linear regression analysis was discussed and quality measures were introduced. Based on regression, ranking schemes for identifying important parameters or variables were discussed. Response surface models based on both regression and interpolation were given. In order to set up response surfaces, support points must be chosen appropriately. Suitable methods for the design of experiments were shown.

Chapter 4

Mechanical vibrations due to random excitations

ABSTRACT: Typical actions on mechanical systems and structures have a considerable temporal variation. This leads to time-dependent dynamic responses. The mathematical framework to describe actions and reactions is based on random process theory.

This chapter starts with the basic description of random processes in both time and frequency domains, including the mechanical transfer functions. Methods to compute the response statistics in stationary and non-stationary situations are described and applied to several problem classes. Analytical methods based on Markov process theory are discussed as well a numerical methods based on Monte Carlo simulation. The chapter concludes with a section on stochastic stability.

4.1 Basic definitions

A random process $X(t)$ is the ensemble of all possible realizations (sample functions) $X(t,\sigma)$ as shown in Fig. 4.1. Here t denotes the independent variable (usually identified with time) and σ denotes chance (randomness).

For any given value of t, $X(t)$ is a random variable. By taking ensemble averages at a fixed value of t, we can define expected values of the random process $X(t)$. These expectations are first of all the mean value function

$$\bar{X}(t) = \mathbf{E}[X(t)] \tag{4.1}$$

and the auto-covariance function

$$R_{XX}(t,s) = \mathbf{E}[(X(t) - \bar{X}(t))(X(s) - \bar{X}(s))] \tag{4.2}$$

From Eq. (4.2) we obtain as a special case for $t = s$

$$R_{XX}(t,t) = \mathbf{E}[(X(t) - \bar{X}(t))^2] = \sigma_X^2(t) \tag{4.3}$$

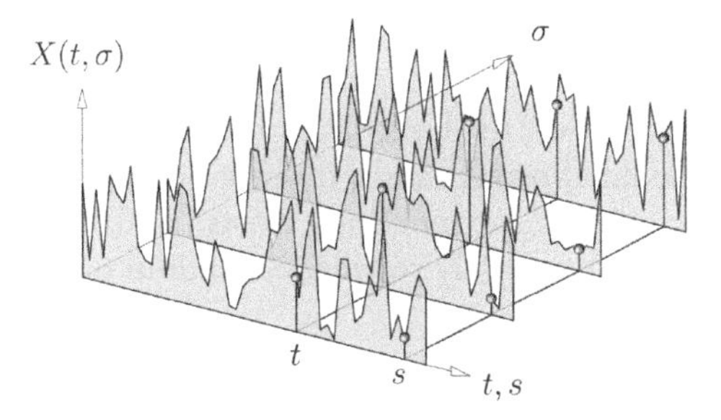

Figure 4.1 Ensemble of realizations of a random process.

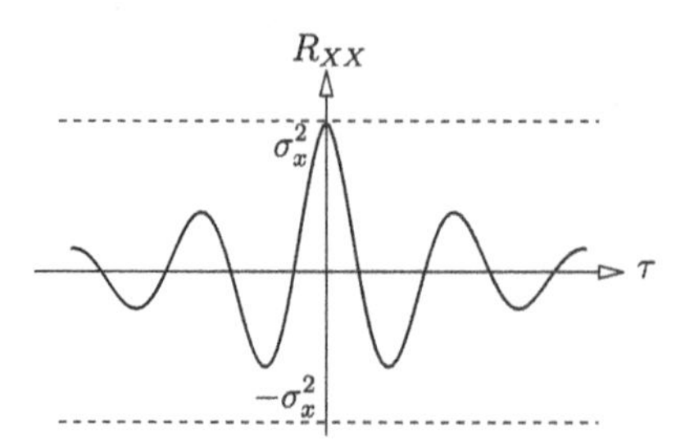

Figure 4.2 Auto-covariance function of a weakly stationary random process.

An enhanced description of a random process involves the probability distribution functions. This includes the one-time distribution function

$$F_X(x,t) = \mathbf{Prob}[X(t) < x] \tag{4.4}$$

and, moreover, all multi-time distribution functions

$$\begin{aligned} &F_X(x_1,t_1;x_2,t_2;\ldots;x_n,t_n) \\ &\quad = \mathbf{Prob}[(X(t_1) < x_1) \wedge (X(t_2) < x_2) \wedge \ldots \wedge (X(t_n) < x_n)] \end{aligned} \tag{4.5}$$

for arbitrary $n \in \mathbb{N}$. If all these distribution functions are (multidimensional) Gaussian distributions, then the process $X(t)$ is called *Gaussian process*. This class of random processes has received particular attention in stochastic dynamics since its properties are easily described in terms of the mean value function and the auto-covariance function only.

A random process is called *weakly stationary* if its mean value function $\bar{X}(t)$ and auto-covariance function $R_{XX}(t,s)$ satisfy the relations

$$\begin{aligned} \bar{X}(t) &= \bar{X} = const. \\ R_{XX}(t,s) &= R_{XX}(s-t) = R_{XX}(\tau) \end{aligned} \tag{4.6}$$

For weakly stationary processes we have

$$\begin{aligned} R_{XX}(\tau) &= R_{XX}(-\tau) \\ \max_{\tau\in\mathbb{R}} |R_{XX}(\tau)| &= R_{XX}(0) = \sigma_X^2 \end{aligned} \tag{4.7}$$

Intuitively, one may expect that for large time separation (i.e. for $\tau \to \pm\infty$) the auto-covariance function should approach zero. If this is actually the case, then the Fourier transform of the auto-covariance functions exists, and we define the *auto-power spectral density* $S_{XX}(\omega)$ of the weakly stationary random process $X(t)$ in terms of

$$S_{XX}(\omega) = \frac{1}{2\pi}\int_{-\infty}^{\infty} R_{XX}(\tau)e^{i\omega\tau}\mathrm{d}\tau \tag{4.8}$$

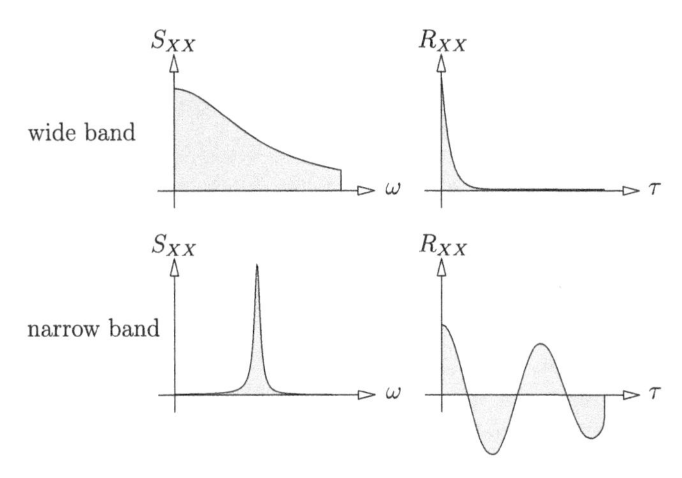

Figure 4.3 Qualitative relation beween PSD and auto-covariance functions for wide band and narrow band random processes.

By inverting this transformation we can recover the auto-covariance function in terms of

$$R_{XX}(\tau) = \int_{-\infty}^{\infty} S_{XX}(\omega) e^{-i\omega\tau} \mathrm{d}\omega \tag{4.9}$$

These equations are frequently called Wiener-Khintchine-relations. Specifically, for $\omega = 0$ we obtain from the previous equation

$$\sigma_X^2 = R_{XX}(0) = \int_{-\infty}^{\infty} S_{XX}(\omega) \mathrm{d}\omega \tag{4.10}$$

This leads to the interpretation of the power spectral density (PSD) as the distribution of the variance of a process over the frequency axis. It forms the basis of the so-called power spectral method of random vibration analysis.

According to the range of frequencies covered by the PSD, the extreme cases of wide-band and narrow band random processes may be distinguished. The qualitative relation between the PSD and the respective auto-covariance functions is shown in Fig. 4.3.

4.2 Markov processes

A continuous Markov process is defined in terms of conditional probability density functions, i.e.

$$f_X(x_n, t_n | x_{n-1}, t_{n-1}; \ldots x_1, t_1) = f_X(x_n, t_n | x_{n-1}, t_{n-1}); \; t_n > t_{n-1} > \cdots > t_1 \tag{4.11}$$

This states that the probability density function of a Markov process at time t_n, given its state at previous times $t_n, t_{n-1} \ldots t_1$, depends only on the value x_{n-1} at time t_{n-1}, i.e. only on the immediate past. As such, a Markov process is called a "one-step-memory random process" (Lin 1976). Using properties of the correlation coefficient function

$$\rho_{XX}(t,s) = \frac{R_{XX}(t,s)}{\sigma_X(t)\sigma_X(s)} \tag{4.12}$$

there is a weaker form of defining a Markov process (Markov process in the wide sense) by

$$\rho_{XX}(t,s) = \rho_{XX}(t,u) \cdot \rho_{XX}(u,s); \; t \leq u \leq s \tag{4.13}$$

If a random process is both wide-sense Markovian and weakly stationary, then

$$\rho_{XX}(t-s) = \rho_{XX}(t-u) \cdot \rho_{XX}(u-s); \; t \leq u \leq s \tag{4.14}$$

or equivalently

$$\rho_{XX}(\tau) = \rho_{XX}(\varphi)\rho_{XX}(\tau-\varphi) \tag{4.15}$$

Taking derivatives with respect to τ we get

$$\dot{\rho}_{XX}(\tau) = \rho_{XX}(\varphi)\dot{\rho}_{XX}(\tau-\varphi) \tag{4.16}$$

and upon setting $\tau = \varphi$

$$\dot{\rho}_{XX}(\tau) = \rho_{XX}(\tau)\dot{\rho}_{XX}(0) = -\beta\rho_{XX}(\tau) \tag{4.17}$$

which due to $\rho_{XX}(0) = 1$ has the unique solution

$$\rho_{XX}(\tau) = \exp(-\beta\tau); \; \beta > 0 \tag{4.18}$$

Hence the auto-covariance function of a weakly stationary wide-sense Markov process is of the exponential type:

$$R_{XX}(\tau) = \sigma_X^2 \exp(-\beta\tau) \tag{4.19}$$

This can easily be Fourier-transformed to give the power spectral density as a rational function

$$S_{XX}(\omega) = \frac{\beta\,\sigma_X^2}{\pi(\beta^2+\omega^2)} \tag{4.20}$$

The `maxima`-code carrying out the integration is given in Listing 4.1.

```
1 assume(beta>0);
2 assume(omega>0);
3 r(t):=S*exp(-beta*t);
4 h:1/2/%pi*2*integrate(r(u)*exp(-%i*omega*u),u,0,inf);
5 s:realpart(h);
6 tex(%);
```

Listing 4.1 Computation of spectral density for Markov process.

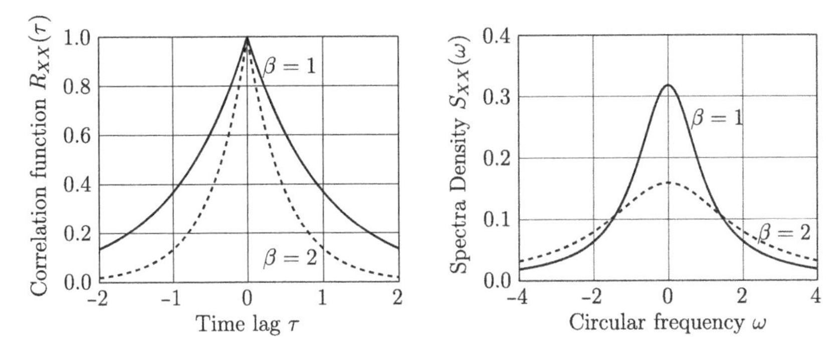

Figure 4.4 Correlation function and power spectral density function of Markov process.

The relation between these functions is shown in Fig. 4.4 for $\sigma_X^2 = 1$ and for different numerical values of β.

The concept of Markov processes can be extended to vector-valued random processes $\mathbf{X}(t)$. In this case, the defining equation becomes

$$f_{\mathbf{X}}(\mathbf{x}_n, t_n | \mathbf{x}_{n-1}, t_{n-1}; \ldots \mathbf{x}_1, t_1) = f_{\mathbf{X}}(\mathbf{x}_n, t_n | \mathbf{x}_{n-1}, t_{n-1}); \; t_n > t_{n-1} > \cdots > t_1 \tag{4.21}$$

The matrix of correlation coefficients $\boldsymbol{\rho}(\tau)$ consequently has the property

$$\boldsymbol{\rho}(t-s) = \boldsymbol{\rho}(t-u)\boldsymbol{\rho}(u-s) \tag{4.22}$$

which implies (cf. the argumentation following Eq. 4.14, Lin 1976)

$$\boldsymbol{\rho}(\tau) = \begin{cases} \exp(-\mathbf{Q}\tau); & \tau > 0 \\ \exp(\mathbf{Q}^T\tau); & \tau < 0 \end{cases} \tag{4.23}$$

in which $\mathbf{Q}$ is a constant matrix whose eigenvalues have positive real parts. Note that the matrix exponential typically contains both exponential and trigonometric functions, thus the correlation coefficients of Markov vector processes will usually be oscillating functions of the time lag τ.

4.2.1 *Upcrossing rates*

For the design of a structure or structural element it is essential to deal with extreme responses to dynamic loading. This means that the probability of exceeding large,

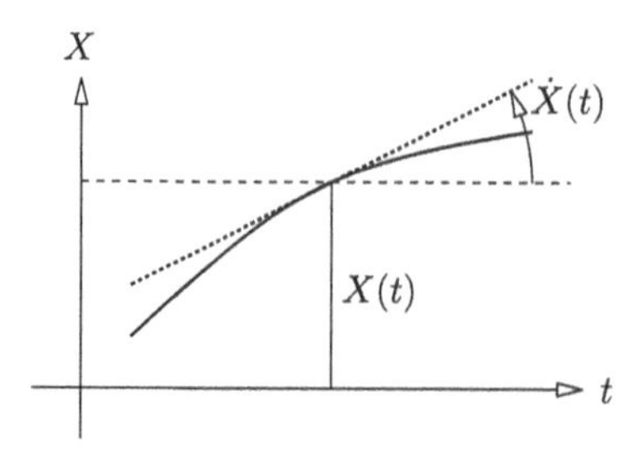

Figure 4.5 Upcrossing event of a random process process.

possibly dangerous levels ξ of the response $X(t)$ should be kept small. Different types of failure can be associated with the exceedance of the threshold value. One possible failure mechanism is sudden failure once the threshold is crossed (ultimate load failure), another possibility is the accumulation of damage due to repeated exceedance of the threshold (fatigue failure). For both types of failure it is essential to investigate the average number of crossings over the threshold per unit time (the upcrossing rate ν_ξ). Upcrossing is an event at time t in which the value of the random process X is below the threshold immediately before t and above the threshold immediately after t. The event occurs while the time-derivative $\dot{X}$ is positive (cf. Fig. 4.5). In order to derive an expression for ν_ξ it is necessary to know the joint probability density function of the random process and its time derivative at any time t, i.e. $f_{X,\dot{X}}(x,\dot{x},t)$. If $\dot{X}$ exists in the mean square sense, then its mean value is zero

$$\mathbf{E}[\dot{X}] = \lim_{\Delta t \to 0} \frac{1}{\Delta t} \mathbf{E}[X(t+\Delta t) - X(t)] = \dot{\bar{X}} = 0 \tag{4.24}$$

and its covariance function is given by

$$\begin{aligned} R_{\dot{X}\dot{X}} = \mathbf{E}[\dot{X}(t)\dot{X}(s)] &= \lim_{u\to 0}\lim_{v\to 0} \mathbf{E}\left[\frac{X(t+u)-X(t)}{u}\frac{X(s+v)-X(s)}{v}\right] \\ &= \lim_{v\to 0}\frac{1}{v}\lim_{u\to 0}\frac{R_{XX}(t+u-s-v)-R_{XX}(t-s-v)}{u} \\ &\quad - \lim_{v\to 0}\frac{1}{v}\lim_{u\to 0}\frac{R_{XX}(t+u-s)+R_{XX}(t-s)}{u} \\ &= \lim_{v\to 0}\frac{1}{v}\left[R'_{XX}(t-s-v) - R'_{XX}(t-s)\right] = -R''_{XX}(t-s) \end{aligned} \tag{4.25}$$

This means that the differentiability of a random process in the mean square sense requires that its auto-covariance function is twice differentiable. This shows that a scalar Markov process is not mean-square differentiable. However, individual components of a vector Markov process may be mean-square differentiable. In a stationary process, the process itself and its time derivative are uncorrelated if taken at the same time t. This is readily shown by taking

$$\mathbf{E}[X(t)\dot{X}(t)] = \lim_{\Delta t \to 0} \frac{1}{\Delta t} \mathbf{E}[X(t)\,(X(t+\Delta t) - X(t))\,]$$
$$= \lim_{\Delta t \to 0} \frac{1}{\Delta t} (R_{XX}(\Delta t) - R_{XX}(0)) = R'_{XX}(0) \tag{4.26}$$

Due to the required symmetry of the auto-covariance function $R_{XX}(\tau)$ with respect to the time separation argument τ, its derivative R'_{XX} is either zero or it does not exist. However, since the existence of the derivative requires the differentiability of R_{XX} we conclude that $X(t)$ and $\dot{X}(t)$ are uncorrelated.

In the case of a Gaussian random process this implies that $X(t)$ and $\dot{X}(t)$ are independent when evaluated at the same time. The joint probability density function of the process and its time derivative is then simply

$$f_{X\dot{X}}(x,\dot{x}) = \frac{1}{2\pi} \frac{1}{\sigma_X \sigma_{\dot{X}}} \exp\left[-\frac{(x-\bar{X})^2}{2\sigma_X^2}\right] \exp\left[-\frac{\dot{x}^2}{2\sigma_{\dot{X}}^2}\right] \tag{4.27}$$

The upcrossing rate ν_ξ of a random process $X(t)$ over a threshold ξ from below can be computed as (Lin 1976):

$$\nu_\xi = \int_0^\infty \dot{x} f_{X\dot{X}}(\xi,\dot{x}) \mathrm{d}\dot{x} \tag{4.28}$$

For a Gaussian process as defined in Eq. 4.27, this evaluates to

$$\nu_\xi = \frac{1}{2\pi} \frac{\sigma_{\dot{X}}}{\sigma_X} \exp\left[-\frac{(x-\bar{X})^2}{2\sigma_X^2}\right] \tag{4.29}$$

By studying the joint probability density function of the process $X(t)$ and its first and second derivatives $\dot{X}(t)$ and $\ddot{X}(t)$, an expression for the probability density function $f_A(a)$ of the peaks A of a random process can be obtained (Lin 1976). For the limiting case of a narrow-band process, $f_A(a)$ becomes

$$f_A(a) = \frac{a}{\sigma_X^2} \exp\left[-\frac{(a-\bar{X})^2}{2\sigma_X^2}\right], \quad a \geq \bar{X} \tag{4.30}$$

4.3 Single-degree-of-freedom system response

4.3.1 *Mean and variance of response*

Consider a mechanical system consisting of a mass m, a viscous damper c and an elastic spring k as shown in Fig. 4.6. The equation of motion (dynamic equilibrium condition) for this system is

$$m\ddot{X} + c\dot{X} + kX = F(t) \tag{4.31}$$

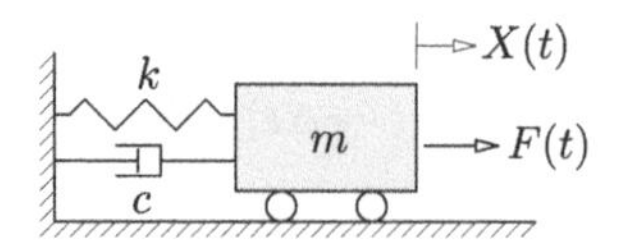

Figure 4.6 Single-degree-of-freedom oscillator.

For this system, we can derive the natural circular frequency ω_0 and the damping ratio ζ as

$$\omega_0 = \sqrt{\frac{k}{m}}; \quad \zeta = \frac{c}{2\sqrt{mk}} \tag{4.32}$$

We assume that the load $F(t)$ acting on the system is a random process. At present we specifically assume that $F(t)$ is a weakly stationary process with a given mean value $\bar{F}$ and a given autocovariance function $R_{FF}(\tau)$. We want to compute the statistical properties of the displacement response $X(t)$, which will be a random process, too. From structural dynamics, we can apply the so-called Duhamel's integral:

$$X(t) = \int_0^t h(t-w)F(w)\mathrm{d}w \tag{4.33}$$

In this equation, $h(u)$ denotes the impulse response function given by

$$h(u) = \begin{cases} \frac{1}{m\omega'} \exp(-\zeta\omega_0 u) \sin \omega' u & u \geq 0 \\ 0 & u < 0 \end{cases} \tag{4.34}$$

with $\omega' = \omega_0\sqrt{1-\zeta^2}$. Applying the expectation operator on Eq. (4.33), we obtain

$$\begin{aligned} \mathbf{E}[X(t)] = \bar{X}(t) &= \mathbf{E}\left[\int_0^t h(t-w)F(w)\mathrm{d}w\right] \\ &= \int_0^t h(t-w)\mathbf{E}[F(w)]\mathrm{d}w = \bar{F}\int_0^t h(t-w)\mathrm{d}w \end{aligned} \tag{4.35}$$

By substituting the variable $u = t - w$ we immediately get

$$\bar{X}(t) = \bar{F}\int_0^t h(u)\mathrm{d}u = \frac{\bar{F}}{m\omega_0^2}\left[1 - \exp(-\zeta\omega_0 t)\left(\zeta \sin \omega' t + \sqrt{1-\zeta^2}\cos \omega' t\right)\right] \tag{4.36}$$

From this it is easily seen that in the limit as $t \to \infty$, we obtain the static solution as the stationary solution

$$\lim_{t\to\infty} \bar{X}(t) = \frac{\bar{F}}{k} = \bar{X}_\infty \tag{4.37}$$

If the damping ratio is not too small, the limit is approached quite rapidly. For numerical values of $\omega_0 = 1$ and $\zeta = 0.05$, this is shown in Fig. 4.7. There is an initial overshoot

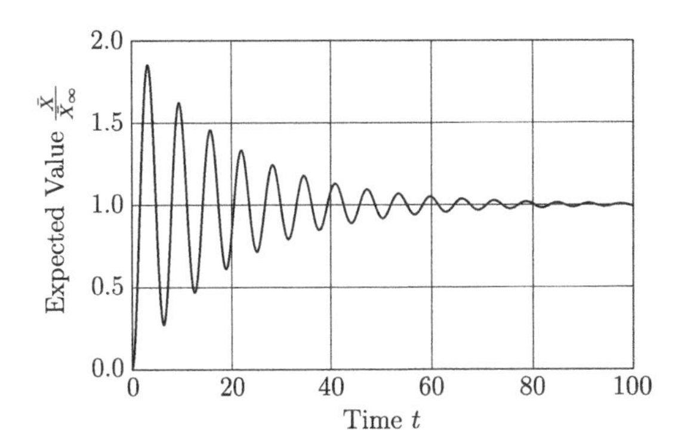

Figure 4.7 Expected displacement response for SDOF systems.

which in structural analysis is commonly called dynamic load factor. It is less or equal to 2. For processes with a sufficiently long duration, the stationary limit $\bar{X}_\infty$ is of main interest. This so-called stationary mean response can also be obtained for finite values of t by assuming that the excitation started in the infinite past. Based on this, Duhamel's integral can be written as

$$X(t) = \int_{-\infty}^{t} h(t-w)F(w)\mathrm{d}w \tag{4.38}$$

Actually, due to the fact that $h(u)=0$ for $u<0$, we can write just as well

$$X(t) = \int_{-\infty}^{\infty} h(t-w)F(w)\mathrm{d}w \tag{4.39}$$

From this, we can easily get

$$\mathbf{E}[X(t)] = \bar{X}(t) = \mathbf{E}\left[\int_{-\infty}^{\infty} h(t-w)F(w)\mathrm{d}w\right] = \bar{F}\int_{0}^{\infty} h(u)\mathrm{d}u = \frac{\bar{F}}{k} \tag{4.40}$$

In the following, we assume that the excitation has been acting since the infinite past. The autocovariance function $R_{XX}(t,s)$ of the response $X(t)$ can then also be computed from Eq. (4.39):

$$\begin{aligned}\mathbf{E}[X(t)X(s)] &= \mathbf{E}\left[\int_{-\infty}^{\infty} h(t-w)F(w)\mathrm{d}w \cdot \int_{-\infty}^{\infty} h(s-z)F(z)\mathrm{d}z\right]\\ &= \mathbf{E}\left[\int_{-\infty}^{\infty}\int_{-\infty}^{\infty} h(t-w)h(s-z)F(w)F(z)\mathrm{d}w\mathrm{d}z\right]\\ &= \int_{-\infty}^{\infty}\int_{-\infty}^{\infty} h(t-w)h(s-z)\mathbf{E}[F(w)F(z)]\mathrm{d}w\mathrm{d}z\end{aligned} \tag{4.41}$$

Subtracting the expected values $\bar{F}$ and $\bar{X}$, respectively, this becomes

$$R_{XX}(t,s) = \int_{-\infty}^{\infty}\int_{-\infty}^{\infty} h(t-w)h(s-z)R_{FF}(w,z)\mathrm{d}w\mathrm{d}z \tag{4.42}$$

With the application of the Wiener-Khintchine-relations Eq. (4.8) we obtain the power spectral density function $S_{XX}(\omega)$ of the response

$$S_{XX}(\omega) = \frac{1}{2\pi}\int_{-\infty}^{\infty}\int_{-\infty}^{\infty}\int_{-\infty}^{\infty} h(t-w)h(t+\tau-z)R_{FF}(z-w)\mathrm{e}^{\mathrm{i}\omega\tau}\mathrm{d}w\mathrm{d}z\mathrm{d}\omega \tag{4.43}$$

Using the substitutions

$$u_1 = z - w; \quad u_2 = t - w; \quad u_3 = t + \tau - z \tag{4.44}$$

with the absolute Jacobian determinant $|J|$ of the coordinate transformation

$$|J| = \left|\frac{\partial(u_1,u_2,u_3)}{\partial(z,w,\tau)}\right| = \begin{vmatrix} 1 & -1 & 0 \\ 0 & -1 & 0 \\ -1 & 0 & 1 \end{vmatrix} = |-1| = 1 \tag{4.45}$$

this can be rewritten as

$$S_{XX}(\omega) = \frac{1}{2\pi}\int_{-\infty}^{\infty} R_{FF}(u_1)\mathrm{e}^{\mathrm{i}\omega u_1}\mathrm{d}u_1 \\ \cdot \int_{-\infty}^{\infty} h(u_2)\mathrm{e}^{-\mathrm{i}\omega u_2}\mathrm{d}u_2 \cdot \int_{-\infty}^{\infty} h(u_3)\mathrm{e}^{\mathrm{i}\omega u_3}\mathrm{d}u_3 \tag{4.46}$$

The first line on the right hand side of this equation apparently represents the power spectral density $S_{FF}(\omega)$ of the excitation. The remaining two integrals are the complex transfer function $H(\omega)$ and its complex conjugate $H^*(\omega)$:

$$H(\omega) = \int_{-\infty}^{\infty} h(u)\mathrm{e}^{-\mathrm{i}\omega u}\mathrm{d}u \tag{4.47}$$

So we may conclude that the power spectral density of the response is given by the simple relation

$$S_{XX}(\omega) = S_{FF}(\omega)H(\omega)H^*(\omega) = S_{FF}(\omega)|H(\omega)|^2 \tag{4.48}$$

Evaluation of Eq. (4.47) yields

$$H(\omega) = \frac{1}{k - m\omega^2 + \mathrm{i}c\omega} \tag{4.49}$$

so that

$$S_{XX}(\omega) = S_{FF}(\omega)\frac{1}{k^2 + (c^2 - 2km)\omega^2 + m^2\omega^4} \tag{4.50}$$

Using Eq. (4.10), the variance σ_X^2 of the displacement response can be computed from

$$\sigma_X^2 = \int_{-\infty}^{\infty} S_{FF}(\omega)|H(\omega)|^2 \mathrm{d}\omega \tag{4.51}$$

Example 4.1 (Cantilever subjected to lateral loading)
As an example, consider a simple cantilever structure subjected to random lateral loading (cf. Fig. 4.8). Structural data is $H = 4\,m$, $EI = 3600\ kN/m^2$, $m = 1\,t$. From this, the lateral stiffness is $k = \frac{3EI}{H^3} = 400\,kN/m$. The load model uses a constant mean value $\bar{F}$ and power spectral density

$$S_{FF}(\omega) = \frac{\sigma_F^2 a}{\pi(a^2 + \omega^2)} \tag{4.52}$$

We assume that $\sigma_F = 0.2\bar{F}$ and $a = 12\,rad/s$. The mean response $\bar{X}$ is readily computed to be

$$\bar{X} = \frac{\bar{F}}{k} = 0.0025\bar{F} \tag{4.53}$$

The power spectral densites (on the positive frequency axis) of load and response are shown in Fig. 4.9. Integration over ω from $-\infty$ to ∞ yields the variance of the displacement response

$$\sigma_X^2 = \frac{13}{5560000}\bar{F}^2 = 2.338 \cdot 10^{-6}\bar{F}^2; \quad \rightarrow \sigma_X = 1.529 \cdot 10^{-3}\bar{F} \tag{4.54}$$

The coefficient of variation of the response is 0.61. This clearly indicates the magnification of the randomness due to dynamic effects.

The result for σ_X^2 as shown above is a closed-form solution obtained by `maxima` using the script as shown in Listing 4.2.

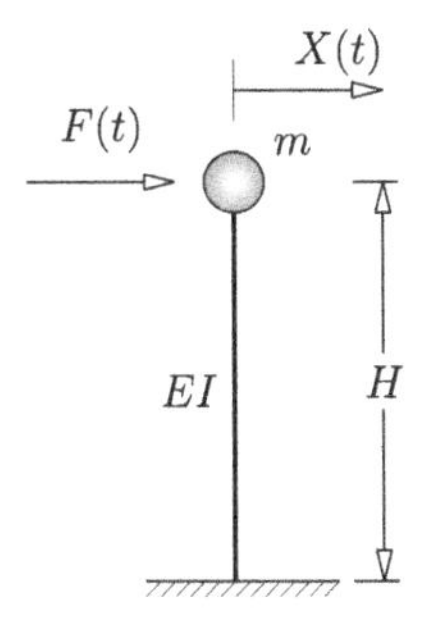

Figure 4.8 Cantilever structure subjected to lateral loading.

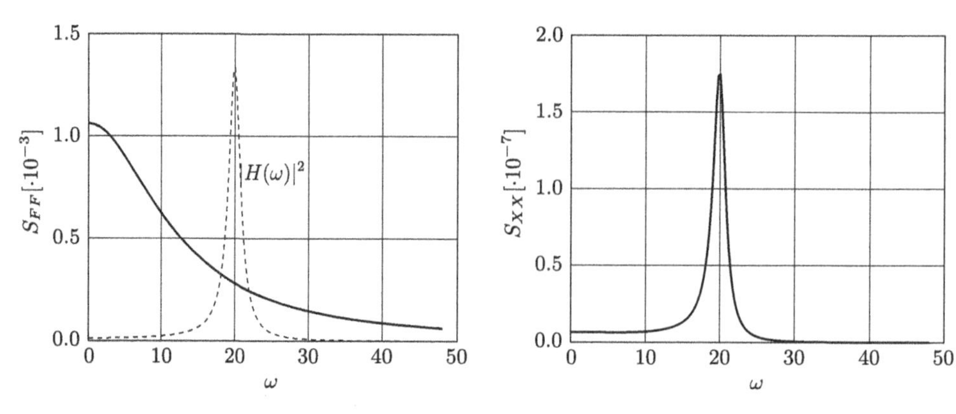

Figure 4.9 Power spectral densities of load and response for cantilever structure.

```
1 sff:1/25*12/%pi/(12^2+om^2);
2 k:400;
3 m:1;
4 c:0.05*sqrt(k*m);
5 hh:1/(k^2+(c^2-2*k*m)*om^2+m^2*om^4);
6 var:integrate(hh*sff,om,-inf,inf);
```

Listing 4.2 Integration for variance of response.

4.3.2 *White noise approximation*

In view of the integral, as given in Eq. (4.51), it is quite obvious that the major contribution to the value of σ_X^2 will most likely come from the frequency range near the natural circular frequency ω_0. Based on this observation, the integral can be approximated by

$$\sigma_x^2 \approx \int_{-\infty}^{\infty} S_{FF}(\omega_0)|H(\omega)|^2 \mathrm{d}\omega = S_{FF}(\omega_0) \int_{-\infty}^{\infty} |H(\omega)|^2 \mathrm{d}\omega \tag{4.55}$$

The integral over the squared magnitude of the complex transfer function can be evaluated in closed form:

$$\int_{-\infty}^{\infty} |H(\omega)|^2 \mathrm{d}\omega = \int_{-\infty}^{\infty} \frac{1}{k^2 + (c^2 - 2km)\omega^2 + m^2\omega^4} \mathrm{d}\omega = \frac{\pi}{kc} \tag{4.56}$$

This approximation procedure can be interpreted as replacing the actual loading process $F(t)$ by another process $W(t)$ which has a constant power spectral density function $S_{WW}(\omega) = const. = S_{FF}(\omega_0)$. Applying this to the previous example with the cantilever under lateral loading, we obtain the approximate result

$$\sigma_X^2 = \frac{3}{1360000}\bar{F}^2 = 2.206 \cdot 10^{-6}\bar{F}^2 \rightarrow \sigma_X = 1.485 \cdot 10^{-3}\bar{F} \tag{4.57}$$

(see the maxima-code in Listing 4.3).

```
1 sff(om):=1/25*12/%pi/(12^2+om^2);
2 k:400;
3 m:1;
4 c:0.05*sqrt(k*m);
5 hh:1/(k^2+(c^2-2*k*m)*om^2+m^2*om^4);
6 white:integrate(hh,om,-inf,inf);
7 om0:20;
8 varapprox:white*sff(om0);
```

Listing 4.3 Integration for white noise.

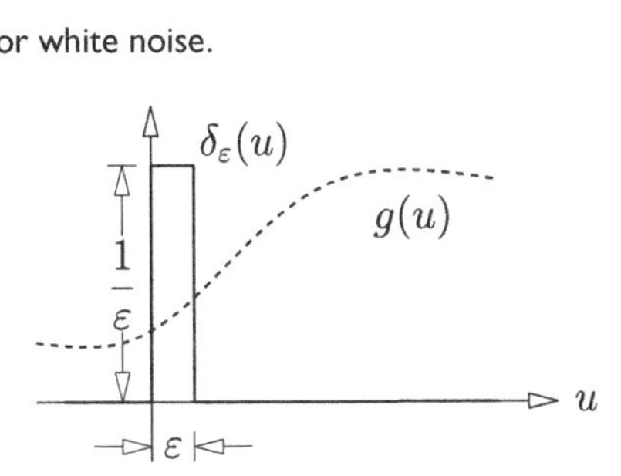

Figure 4.10 Dirac Delta function as limit of rectangular function.

It should be noted, however, that such a process with constant power spectral density cannot exist in reality since according to Eq. (4.10) its variance σ_W^2 should be infinite. Due to the equally distributed frequency components, such a fictitious process is called "white noise" (in analogy to white light containing all visible frequencies in equal intensity). Formally, the autocorrelation function $R_{WW}(\tau)$ of a process with constant power spectral density S_0 can be constructed from Eq. (4.9):

$$R_{WW}(\tau) = \int_{-\infty}^{\infty} S_0(\omega)e^{-i\omega\tau}\mathrm{d}\omega = 2\pi S_0\delta(\tau) \tag{4.58}$$

Here, $\delta(.)$ denotes the so-called Dirac's Delta function with the properties

$$\delta(u) = 0\ \forall u \neq 0; \qquad \int_{-\infty}^{\infty} \delta(u)g(u)\mathrm{d}u = g(0) \tag{4.59}$$

The latter property is true for all functions $g(u)$ which are continuous in a vicinity of $u = 0$. The Delta function can be interpreted for example as the limiting case of a rectangular function $\delta_\varepsilon(u)$ (cf. Fig. 4.10). We define δ_ε as

$$\delta_\varepsilon(u) = \begin{cases} \frac{1}{\varepsilon} & 0 \le u \le \varepsilon \\ 0 & \text{else} \end{cases} \tag{4.60}$$

The function g can be expanded in a Taylor series about $u = 0$

$$g(u) = g(0) + g'(0)u + \frac{1}{2}g''(0)u^2 + \cdots \tag{4.61}$$

```
1 h(t):=1/m/omp*exp(-zeta*omega*t)*sin(omp*t);
2 assume(t>0);
3 x:2*pi*S0*integrate(h(u)^2,u,0,t);
4 x,omp=omega*sqrt(1-zeta^2);
5 factor(%);
6 ratsimp(%);
7 tex(%);
```

Listing 4.4 Integration for transient stochastic response.

so that

$$\int_{-\infty}^{\infty} \delta_\varepsilon(u)g(u)\mathrm{d}u = \int_{-\infty}^{\infty} \delta_\varepsilon(u)\left[g(0)+g'(0)u+\ldots\right]\mathrm{d}u$$
$$= \int_0^{\varepsilon} \frac{1}{\varepsilon}\left[g(0)+g'(0)u+\ldots\right]\mathrm{d}u = \frac{1}{\varepsilon}\left[g(0)\varepsilon+g'(0)\frac{\varepsilon^2}{2}+\ldots\right] \tag{4.62}$$

In the limit as $\varepsilon \to 0$ we obtain

$$\lim_{\varepsilon\to 0}\int_{-\infty}^{\infty} \delta_\varepsilon(u)g(u)\mathrm{d}u = g(0) \tag{4.63}$$

Returning to the expression for the autocovariance function of the response as given in Eq. (4.41), the above property of the Delta function allows the computation of an expression for the time-dependent variance $\sigma_X^2(t)$ of the response to white noise excitation

$$\sigma_X^2(t) = R_{XX}(t,t) = \int_0^t\int_0^t h(t-w)h(t-z)2\pi S_0\delta(z-w)\mathrm{d}w\mathrm{d}z$$
$$= 2\pi S_0\int_0^t h(t-z)^2\mathrm{d}z \tag{4.64}$$

By substituting the variable $u=t-z$ we obtain (see the maxima-script in Listing 4.4)

$$\sigma_X^2(t) = 2\pi S_0\int_0^t h(u)^2\mathrm{d}u$$
$$= \frac{\pi S_0}{kc}\left[1-\exp(-2\zeta\omega_0 t)\left(\frac{\omega_0^2}{\omega'^2}-\frac{\zeta^2\omega_0^2}{\omega'^2}\cos 2\omega' t+\zeta\sin 2\omega' t\right)\right] \tag{4.65}$$

For numerical values of $k=1$ N/m, $m=1$ kg, $\zeta=0.05$, $S_0=1\ \mathrm{N^2s}$, the result of Eq. 4.65 is shown with the label "exact" in Fig. 4.11. From Eq. 4.65 it can be seen that the

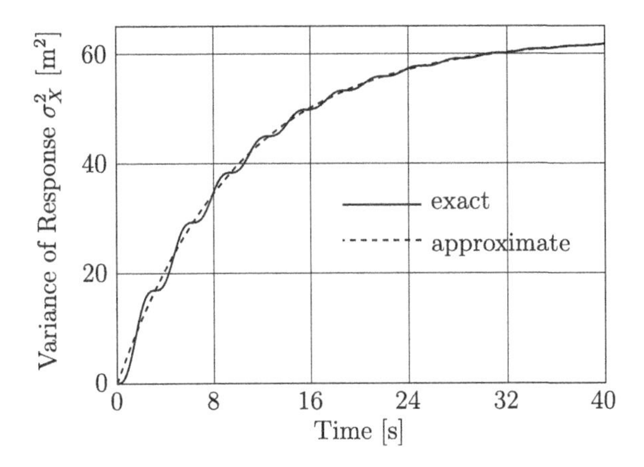

Figure 4.11 Variance of response to white noise excitation.

contributions from the trigonometric functions are of the order of the damping ratio ζ or smaller. Therefore, omitting these terms yields a simple approximation in the form of

$$\sigma_X^2(t) \approx \frac{\pi S_0}{kc}\left[1 - \exp(-2\zeta\omega_0 t)\right] \tag{4.66}$$

The result from this equation is shown with the label “approximate” in Fig. 4.11. We will return to this result in section 4.4.2 on covariance analysis.

4.4 Multi-degree-of-freedom response

4.4.1 *Equations of motion*

For a linear multi-degree-of-freedom system the equations of motion can be written in matrix-vector form as

$$\mathbf{M}\ddot{\mathbf{X}} + \mathbf{C}\dot{\mathbf{X}} + \mathbf{K}\mathbf{X} = \mathbf{F}(t) \tag{4.67}$$

together with appropriate initial conditions for $\mathbf{X}$ and $\dot{\mathbf{X}}$. Here, the vectors $\mathbf{X}$ and $\dot{\mathbf{X}}$ have the dimension n, the symmetric and non-negative matrices $\mathbf{M}$, $\mathbf{C}$ and $\mathbf{K}$ have the size $n \times n$, and $\mathbf{F}(t)$ is an n-dimensional vector valued random process. We assume that at least the second order statistics of $\mathbf{F}$ are known. For an important class of nonlinear structural systems, the equations of motion can be written as

$$\mathbf{M}\ddot{\mathbf{X}} + \mathbf{g}(\mathbf{X}, \dot{\mathbf{X}}) = \mathbf{F}(t) \tag{4.68}$$

The nonlinearity in Eq. 4.68 is present in the function $\mathbf{g}$ involving both restoring forces and damping.

4.4.2 Covariance analysis

In the case of delta-correlated excitation processes, there is a direct way of deriving equations for the covariance matrix of the response vector components. This is especially advantageous for multi-degree-of-freedom systems. We assume that the matrix of auto- and cross-covariance functions of the components of the excitation vector $\mathbf{F}(t)$ are given in the form of

$$\mathbf{R}_{\mathbf{FF}}(t, t+\tau) = \mathbf{D}(t)\delta(\tau) \tag{4.69}$$

Here, $\mathbf{D}$ is an arbitrary cross intensity matrix of the size $n \times n$, possibly depending on time t. This means that the excitation process $\mathbf{F}(t)$ is a multi-dimensional white noise process. Now the equation of motion is represented in phase space, i.e. the response is described in terms of a state vector $\mathbf{Y}$ containing the displacements $\mathbf{X}$ and the velocities $\dot{\mathbf{X}}$. In phase space, the equation of motion becomes (cf. Eq. 4.67):

$$\dot{\mathbf{Y}} - \mathbf{G}\mathbf{Y} = \mathbf{g}(t) \tag{4.70}$$

The $2n \times 2n$-matrix $\mathbf{G}$ is assembled from the mass, stiffness and damping matrices as

$$\mathbf{G} = \begin{bmatrix} 0 & \mathbf{I} \\ -\mathbf{M}^{-1}\mathbf{K} & -\mathbf{M}^{-1}\mathbf{C} \end{bmatrix} \tag{4.71}$$

The covariance matrix $\mathbf{R}_{\mathbf{YY}}$ also has the size of $2n \times 2n$. This matrix satisfies the differential equation (see e.g. Lin 1976)

$$\dot{\mathbf{R}}_{\mathbf{YY}} = \mathbf{G}\mathbf{R}_{\mathbf{YY}} + \mathbf{R}_{\mathbf{YY}}\mathbf{G}^T + \mathbf{B}(t) \tag{4.72}$$

Here, the matrix $\mathbf{B}$ is defined by

$$\mathbf{B} = \begin{bmatrix} 0 & 0 \\ 0 & \mathbf{M}^{-1}\mathbf{D}(t)\mathbf{M}^{-1} \end{bmatrix} \tag{4.73}$$

This equation can be solved, for instance, by arranging the elements of $\mathbf{R}_{\mathbf{YY}}$ into a vector $\mathbf{r}$. Hereby, the symmetry of $\mathbf{R}_{\mathbf{YY}}$ can be utilized to reduce the problem size. The vector $\mathbf{r}$ then contains $\frac{n(n+1)}{2}$ elements. Correspondingly, the coefficients of the matrix $\mathbf{G}$ are then arranged into another matrix $\mathbf{H}$, and the matrix $\mathbf{B}$ is put into a vector $\mathbf{b}$. Thus we obtain the system of linear equations

$$\dot{\mathbf{r}} = \mathbf{H}\mathbf{r} + \mathbf{b}(t) \tag{4.74}$$

which can be solved using standard methods. The covariance matrix $\mathbf{R}_{\mathbf{XX}}$ of the displacements is a sub-matrix of the size $n \times n$.

An important special case is the *stationary state*, i.e. the case in which $\mathbf{B} = \text{const.}$ and $\dot{\mathbf{R}}_{\mathbf{YY}} = 0$. The excitation process in this case possesses a constant power spectral density matrix $\mathbf{S}_0 = \frac{1}{2\pi}\mathbf{B}$. If we write the matrix $\mathbf{R}_{\mathbf{YY}}$ for this case in block notation

$$\mathbf{R}_{\mathbf{YY}} = \begin{bmatrix} \mathbf{R}_{\mathbf{XX}} & \mathbf{R}_{\mathbf{X}\dot{\mathbf{X}}} \\ \mathbf{R}_{\dot{\mathbf{X}}\mathbf{X}} & \mathbf{R}_{\dot{\mathbf{X}}\dot{\mathbf{X}}} \end{bmatrix}; \quad \mathbf{R}_{\dot{\mathbf{X}}\mathbf{X}} = \mathbf{R}_{\mathbf{X}\dot{\mathbf{X}}}^T \tag{4.75}$$

we obtain for the stationary state

$$
\begin{aligned}
&\mathbf{R}_{\mathbf{X}\dot{\mathbf{X}}} + \mathbf{R}_{\dot{\mathbf{X}}\mathbf{X}} = 0 \\
&\mathbf{K}\mathbf{R}_{\mathbf{X}\mathbf{X}} + \mathbf{C}\mathbf{R}_{\dot{\mathbf{X}}\mathbf{X}} - \mathbf{M}\mathbf{R}_{\dot{\mathbf{X}}\dot{\mathbf{X}}} = 0 \\
&\mathbf{R}_{\mathbf{X}\mathbf{X}}\mathbf{K} + \mathbf{R}_{\mathbf{X}\dot{\mathbf{X}}}\mathbf{C} - \mathbf{R}_{\dot{\mathbf{X}}\dot{\mathbf{X}}}\mathbf{M} = 0 \\
&\mathbf{K}(\mathbf{R}_{\mathbf{X}\dot{\mathbf{X}}} + \mathbf{R}_{\dot{\mathbf{X}}\mathbf{X}}) + 2\mathbf{C}\mathbf{R}_{\dot{\mathbf{X}}\dot{\mathbf{X}}} = \mathbf{M}\mathbf{B}
\end{aligned}
\tag{4.76}
$$

From this we can immediately get the covariance matrix of the displacements

$$
\mathbf{R}_{\mathbf{X}\mathbf{X}} = \frac{1}{2}\mathbf{K}^{-1}\mathbf{M}\mathbf{C}^{-1}\mathbf{M}\mathbf{B} \tag{4.77}
$$

Note: For a SDOF-system this reduces to

$$
\sigma_X^2 = \frac{\pi S_0}{kc}; \quad \sigma_{\dot{X}}^2 = \frac{\pi S_0}{mc} \tag{4.78}
$$

The effort required for the numerical solution can be reduced substantially if modal damping is assumed and approximate validity of some of the relations in Eq. 4.76 is assumed as well (Bucher 1988b). For this case, we can obtain decoupled approximate equations for the elements R_{ij} of the modal covariance matrix

$$
a_{ij}\dot{R}_{ij} + b_{ij}R_{ij} = B_{ij}(t) \tag{4.79}
$$

in which the constants a_{ij} and b_{ij} are given by

$$
\begin{aligned}
a_{ij} &= \omega_i\omega_j\frac{\zeta_i\omega_j + \zeta_j\omega_i}{\zeta_i\omega_i + \zeta_j\omega_j} + \frac{\omega_i^2 + \omega_j^2}{2} \\
b_{ij} &= \omega_i\omega_j\zeta_i\omega_j + \zeta_j\omega_i\zeta_i\omega_i + \zeta_j\omega_j\left(\frac{\omega_i}{4\zeta_i} + \frac{\omega_j}{4\zeta_j} - \frac{\omega_i^2}{4\zeta_j\omega_j} - \frac{\omega_j^2}{4\zeta_i\omega_i} - 2\zeta_i\omega_i - 2\zeta_j\omega_j\right) \\
&\quad + \frac{\omega_i^2\omega_j}{4\zeta_j} + \frac{\omega_i\omega_j^2}{4\zeta_i} - \frac{\omega_i^3}{4\zeta_i} - \frac{\omega_j^3}{4\zeta_j}
\end{aligned}
\tag{4.80}
$$

This leads directly to the modal covariance matrix. For the case $i=j$, Eq. 4.79 reduces to

$$
\dot{R}_{ii} + 2\zeta_i\omega_i R_{ii} = \frac{B_{ii}}{2\omega_i^2} \tag{4.81}
$$

Filtered white noise excitation

The limitation to delta-correlated processes (i.e. generalized white noise) can be lifted by introducing filters. In this approach, the output responses of linear filters to white noise are applied as loads to the structure. A well-known example from earthquake engineering is the Kanai-Tajimi filter. Here, the ground acceleration $a(t)$ is defined as a linear combination of the displacement and velocity of a system with a single degree of freedom. This SDOF system is characterized by its natural circular frequency ω_g and

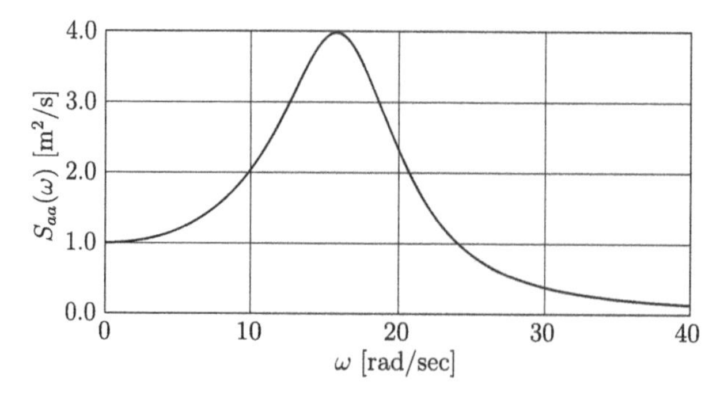

Figure 4.12 Power spectral density of ground acceleration using the Kanai/Tajimi model.

the damping ratio ζ_g. The input $w(t)$ to this filter system is white noise with a power spectral density S_0. The equation of motion for the filter system is

$$\ddot{z} + 2\zeta_g\omega_g\dot{z} + \omega_g^2 z = w(t) \tag{4.82}$$

The ground acceleration process $a(t)$ is then defined as

$$a(t) = 2\zeta_g\omega_g\dot{z} + \omega_g^2 z \tag{4.83}$$

It can be seen that the power spectral density $S_{aa}(\omega)$ has a significant frequency dependence

$$S_{aa}(\omega) = S_0 \frac{4\zeta_g^2\omega_g^2\omega^2 + \omega_g^4}{(\omega_g^2 - \omega^2)^2 + 4\zeta_g^2\omega_g^2\omega^2} \tag{4.84}$$

In Fig. 4.12 this function is plotted for numerical values of $S_0 = 1\,\mathrm{m^2/s}$, $\omega_g = 17\,\mathrm{rad/s}$ and $\zeta_g = 0.3$. Fig. 4.12 clearly shows that significant contributions are present from the frequency range near $\omega = 0$. In a nonlinear analysis these low-frequency components may lead to excessively large plastic drift. In order to avoid this, the Kanai-Tajimi model can be modified such that low-frequency components are reduced (Clough and Penzien 1993).

Example 4.2 (Covariance analysis for SDOF-system)
For such a system the covariance matrix of the state vector contains only four (three mutually different) entries

$$\mathbf{R_{YY}} = \begin{bmatrix} r_{XX} & r_{X\dot{X}} \\ r_{X\dot{X}} & r_{\dot{X}\dot{X}} \end{bmatrix}; \quad \mathbf{G} = \begin{bmatrix} 0 & 1 \\ -\frac{k}{m} & -\frac{c}{m} \end{bmatrix} \tag{4.85}$$

We assume that the excitation is a non-stationary (amplitude modulated) white noise excitation with a power spectral density S_0 and a time envelope $e(t)$. Then, the matrix $\mathbf{B}$ is defined by

$$\mathbf{B} = \begin{bmatrix} 0 & 0 \\ 0 & \frac{2\pi S_0}{m} \end{bmatrix} e^2(t) \tag{4.86}$$

The differential equation 4.72 according to (4.74) can be written as

$$\frac{\mathrm{d}}{\mathrm{d}t}\begin{bmatrix} r_{XX} \\ r_{X\dot{X}} \\ r_{\dot{X}\dot{X}} \end{bmatrix} = \begin{bmatrix} 0 & 2 & 0 \\ -\frac{k}{m} & -\frac{c}{m} & 1 \\ 0 & -\frac{2k}{m} & -\frac{2c}{m} \end{bmatrix}\begin{bmatrix} r_{XX} \\ r_{X\dot{X}} \\ r_{\dot{X}\dot{X}} \end{bmatrix} + \begin{bmatrix} 0 \\ 0 \\ \frac{2\pi S_0}{m} e^2(t) \end{bmatrix} \tag{4.87}$$

which can be written in the symbolic form

$$\dot{\mathbf{r}} = \mathbf{H}\mathbf{r} + \mathbf{h}(t) \tag{4.88}$$

with the initial condition $\mathbf{r}(0) = \mathbf{0}$. The general solution to this equation can be obtained by quadrature

$$\mathbf{r}(t) = \exp(\mathbf{H}t)\mathbf{r}(0) + \int_0^t \exp[\mathbf{H}(t-\tau)]\mathbf{h}(\tau)\mathrm{d}\tau \tag{4.89}$$

which, for $\mathbf{h}$ constant in a time interval Δt, evaluates to

$$\mathbf{r}(\Delta t) = \exp(\mathbf{H}\Delta t)\mathbf{r}(0) + [\exp(\mathbf{H}\Delta t) - \mathbf{I}]\mathbf{H}^{-1}\mathbf{h} \tag{4.90}$$

Note that the matrix exponential exp $(\mathbf{A})$ is readily computed by means of diagonalizing the matrix $\mathbf{A}$ such that

$$\mathbf{A} = \mathbf{T}\mathbf{\Lambda}\mathbf{T}^{-1} \tag{4.91}$$

in which $\mathbf{\Lambda}$ is a diagonal matrix containing the (possibly complex) eigenvalues λ_i of $\mathbf{A}$ and $\mathbf{T}$ is the matrix of corresponding right eigenvectors. This step is always possible if the eigenvalues are distinct. Using Eq. 4.91, the matrix exponential can be computed using the standard series expansion for exp (.), i.e.

$$\begin{aligned} \exp(\mathbf{A}) &= \sum_{k=0}^{\infty} \frac{1}{k!}\mathbf{A}^k = \mathbf{I} + \mathbf{T}\mathbf{\Lambda}\mathbf{T}^{-1} + \frac{1}{2}\mathbf{T}\mathbf{\Lambda}\mathbf{T}^{-1}\mathbf{T}\mathbf{\Lambda}\mathbf{T}^{-1} + \cdots \\ &= \mathbf{T}\mathbf{T}^{-1} + \mathbf{T}\mathbf{\Lambda}\mathbf{T}^{-1} + \frac{1}{2}\mathbf{T}\mathbf{\Lambda}^2\mathbf{T}^{-1} + \cdots \\ &= \mathbf{T}\left(\mathbf{I} + \mathbf{\Lambda} + \frac{1}{2}\mathbf{\Lambda}^2 + \cdots\right)\mathbf{T}^{-1} = \mathbf{T}\exp(\mathbf{\Lambda})\mathbf{T}^{-1} \end{aligned} \tag{4.92}$$

The matrix exponential of the diagonal matrix $\mathbf{\Lambda}$ is simply a diagonal matrix containing the exponentials of the eigenvalues λ_i.

The time-dependent intensity (envelope) is assumed to be

$$e(t) = 4 \cdot [\exp(-0.25t) - \exp(-0.50t)] \tag{4.93}$$

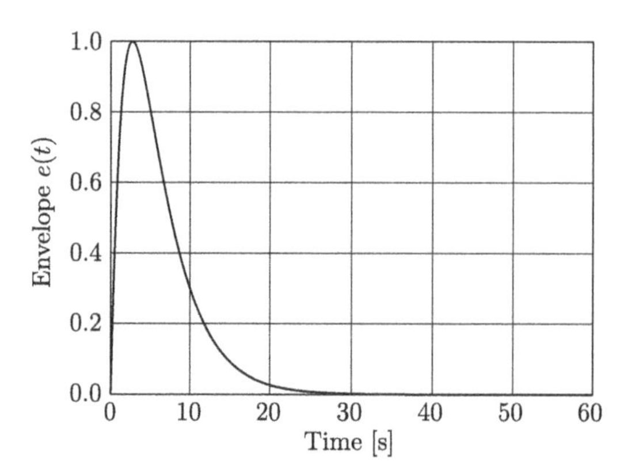

Figure 4.13 Time-dependent amplitude-modulating function (envelope).

(cf. Fig. 4.13). The exact solution from the differential equation (4.72) is obtained using the octave-script as shown in Listing 4.5. The numerical results for $\sigma_X(t)$ are shown in Fig. 4.14.

Upon application of the approximations as mentioned above we obtain the first-order differential equation

$$\frac{\mathrm{d}}{\mathrm{d}t}\sigma_X^2 = -\frac{c}{m}\sigma_X^2 + \frac{S_0 \pi m}{k} e^2(t) \tag{4.94}$$

The solution to this equation is also shown in Fig. 4.14. It can be seen that the approximate solution does not contain oscillations. However, apart from that it matches the exact solution quite well.

Exercise 4.1 (Transient stochastic response)
Compute the variance of the displacement response for a system as defined in the previous example to a nonstationary white noise with an amplitude-modulating function

$$e(t) = \begin{cases} 1; & 0 \leq t \leq T \\ 0; & \text{else} \end{cases} \tag{4.95}$$

for the time interval $[0, 3T]$ with $T = 20$.

Solution: The resulting standard deviation $\sigma_X(t)$ is shown in Fig. 4.15.

Example 4.3 (Stationary response of SDOF oscillator to Kanai-Tajimi excitation)
Consider an SDOF-system with mass m, viscous damping c and stiffness k subjected

```
1  function env=env(t)
2  env=4*(exp(-t/4)-exp(-t/2));
3  endfunction
4  %
5
6  k = 1;
7  m = 1;
8  c = 0.02;
9  S0 = 0.003;
10
11 h=[0;0;2*pi*S0/m];
12 H=[0,2,0;
13 -k/m,-c/m,1;
14 0,-2*k/m,-2*c/m];
15
16 tu = 0;
17 to = 60;
18 NT = 300;
19 dt = (to-tu)/NT;
20
21 [T,L] = eig(H);
22 ee=zeros(3,3);
23 for k=1:3
24   ee(k,k) = exp(L(k,k)*dt);
25   endfor
26 expt = T*ee*inv(T);
27 mat2 = (expt - eye(3))*inv(H);
28 r0=[0;0;0];
29
30 fid = fopen("instat2.txt", "w");
31 fprintf(fid, "%g %g\n", tu, 0);
32 for i=1:NT
33   r1=expt*r0 + mat2*h*env((i-.5)*dt)^2;
34   fprintf(fid, "%g %g\n", dt*i, sqrt(r1(1)));
35   r0 = r1;
36   endfor
37 fclose(fid);
```

Listing 4.5 Computation of transient stochastic response.

to a ground acceleration of the Kanai-Tajimi type. The linear filter representing the ground is driven by stationary white noise with spectral density $S_0 = \frac{1}{2\pi}$. The coupled system of filter and oscillator is described by four state variables, i.e. the oscillator displacement X and velocity $\dot{X}$ as well as the filter displacement Z and velocity $\dot{Z}$

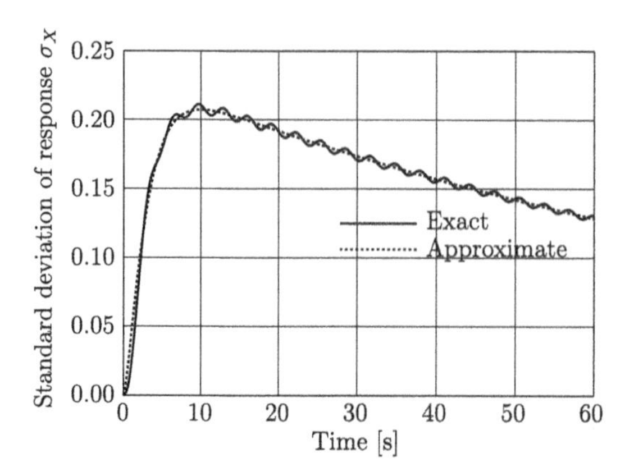

Figure 4.14 Transient standard deviation of the displacement response.

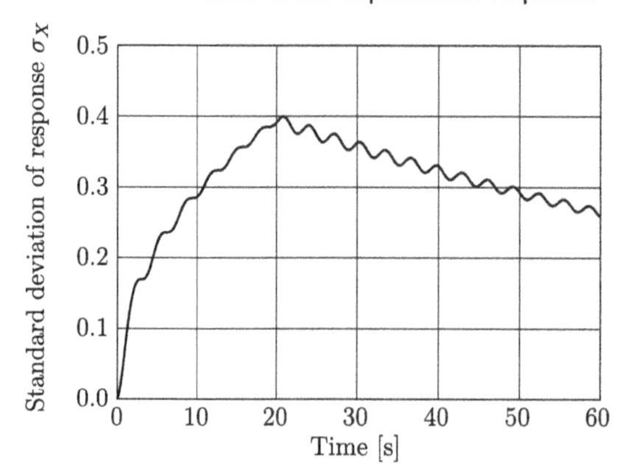

Figure 4.15 Transient standard deviation of the displacement response to pulse modulated white noise.

(cf. Eq. 4.82). The state vector is then $\mathbf{Y} = [X, \dot{X}, Z, \dot{Z}]^T$. It is governed by the system of differential equations

$$\frac{\mathrm{d}}{\mathrm{d}t}\begin{bmatrix} Z \\ \dot{Z} \\ X \\ \dot{X} \end{bmatrix} = \begin{bmatrix} 0 & 1 & 0 & 0 \\ -\omega_g^2 & -2\zeta_g\omega_g & 0 & 0 \\ 0 & 0 & 0 & 1 \\ \omega_g^2 & 2\zeta_g\omega_g & -\omega_0^2 & -2\zeta\omega_0 \end{bmatrix}\begin{bmatrix} Z \\ \dot{Z} \\ X \\ \dot{X} \end{bmatrix} + \begin{bmatrix} 0 \\ W(t) \\ 0 \\ 0 \end{bmatrix} \tag{4.96}$$

in which $\omega_0^2 = \frac{k}{m}$ and $2\zeta\omega_0 = \frac{c}{m}$. This equation is of the form of Eq. 4.70

$$\dot{\mathbf{Y}} - \mathbf{G}\mathbf{Y} = \mathbf{g}W(t) \tag{4.97}$$

Here, $W(t)$ is white noise with a spectral density S_0. The covariance matrix $\mathbf{R_{YY}}$ is governed by Eq. 4.72 which in the case of stationarity reduces to

$$\mathbf{G}\mathbf{R_{YY}} + \mathbf{R_{YY}}\mathbf{G}^T + \mathbf{B} = 0 \tag{4.98}$$

A direct analytical solution for the variances of the response is given by Wall and Bucher 1987. Kronecker products provide an alternative way of rearranging this system of linear equations to solve for the elements of $\mathbf{R_{YY}}$ (see e.g. Di Paola and Elishakoff 1996). The Kronecker product $\mathbf{K}$ of two matrices $\mathbf{A}$ and $\mathbf{B}$ (for clarity and ease of presentation, we assume here that both matrices have the same size $n \times n$) is defined as

$$\mathbf{K} = \mathbf{A} \otimes \mathbf{B} = \begin{bmatrix} a_{11}\mathbf{B} & \dots & a_{1n}\mathbf{B} \\ \vdots & \ddots & \vdots \\ a_{n1}\mathbf{B} & \dots & a_{nn}\mathbf{B} \end{bmatrix} \tag{4.99}$$

Arranging the elements of a regular matrix product $\mathbf{AXB}$ of three matrices into a vector can be accomplished by

$$\text{vec}(\mathbf{AXB}) = [\mathbf{B}^T \otimes \mathbf{A}]\text{vec}(\mathbf{X}) \tag{4.100}$$

Here vec(.) denotes arranging the elements of an $n \times n$-matrix into a column vector of size n^2. In this way, Eq. 4.98 can be rewritten as

$$[\mathbf{I} \otimes \mathbf{G} + \mathbf{G} \otimes \mathbf{I}]\text{vec}(\mathbf{R_{YY}}) = -\text{vec}(\mathbf{B}) \tag{4.101}$$

which can be solved immediately for vec($\mathbf{R_{YY}}$). Numerical values chosen for this example are $S_0 = 1\,\text{m}^2/\text{s}$, $\omega_g = 17\,\text{rad/s}$ and $\zeta_g = 0.3$ together with structural parameters $\omega_0 = 6\,\text{rad/s}$ and $\zeta = 0.1$. The `octave`-code to carry out this solution is given in Listing 4.6.

```
1  omg=17
2  zetag=0.3
3  om=6
4  zeta=0.1
5  S0=1
6  G=[0,1,0,0;
7  -omg^2, -2*zetag*omg, 0, 0;
8  0,0,0,1;
9  omg^2,2*zetag*omg,-om^2, -2*zeta*om];
10 n=rows(G)
11 I=eye(n,n);
12 GI=kron(I,G)+kron(G,I)
13 B=zeros(n,n);
14 B(2,2) = 2*pi*S0;
15 b=-vec(B)
16 c=GI\b
17 C=[c(1:4),c(5:8),c(9:12),c(13:16)]
```

Listing 4.6 Computation of stationary response using Kronecker products.

The result is

```
1  C =
2      0.00107  -0.00000  -0.00009   0.02508
3     -0.00000   0.30800  -0.02508  -0.28104
4     -0.00009  -0.02508   0.09362   0.00000
5      0.02508  -0.28104  -0.00000   3.65147
```

4.4.3 *First passage probability*

In the probability-based design of structures there frequently is a limiting threshold of response quantities such as displacements or stresses which should not be exceeded within a given time period T. Of course, these non-exceedance events have to be characterized in probabilistic terms. There is extensive literature on the topic of "first passage probability". For an overview, see e.g. Macke and Bucher 2003.

For structures with low damping the response to stochastic excitation is typically narrow-banded and contains significant contributions to the variance mainly in the vicinity of the fundamental natural frequency. Thus considerations may be limited to the response in the fundamental mode of vibration. Let the critical magnitude of the response $X(t)$ (e.g. of a displacement) be defined by a threshold value ξ (c. Fig. 4.16). If the threshold is exceeded then we expect a critical or unsafe state of the structure. The structural design must guarantee that this critical state is reached only with a small probability. This probability obviously depends on the magnitude of he threshold ξ. In addition to that, the statistical properties of X and the observation interval T play an important role. We will now derive a simple approximation fo the first passage probability $P_E(\xi, T)$.

First passage failure occurs if $X(t)$ exceeds the threshold ξ at least once within the time interval $[0, T]$

$$P_E(\xi, T) = \textbf{Prob}[\max_{t \in [0,T]} X(t) \geq \xi] \tag{4.102}$$

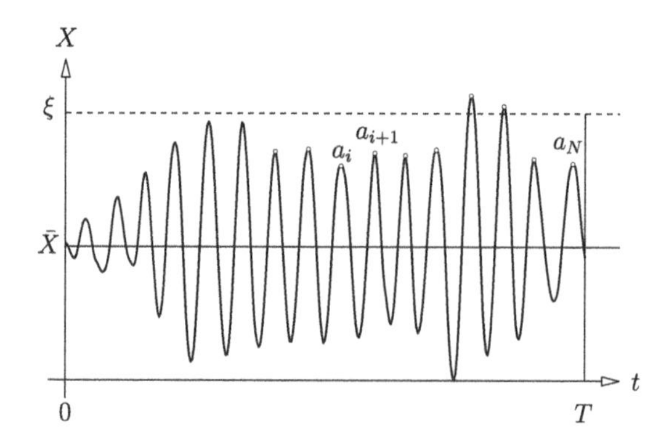

Figure 4.16 Narrow-banded stochastic response.

Since $X(t)$ a continuous process this condition can be replaced by the condition that the largest maximum exceeds the threshold

$$P_E(\xi, T) = \mathbf{Prob}[\max_{i=1,\ldots N} a_i \geq \xi] \tag{4.103}$$

This, in turn, is equivalent to the condition that at least one local maximum exceeds the threshold, i.e. the first, or the second, or...

$$P_E(\xi, T) = \mathbf{Prob}[(a_1 \geq \xi) \vee (a_2 \geq \xi) \vee \ldots (a_N \geq \xi)] \tag{4.104}$$

The complementary event, the non-exceedance of the threshold, has the probability

$$1 - P_E(\xi, T) = \mathbf{Prob}[(a_1 < \xi) \wedge (a_2 < \xi) \wedge \ldots (a_N < \xi)] \tag{4.105}$$

At this point we postulated that the individual events $a_i < \xi$ are statistically independent of each other. It can be shown that this assumption is at least asymptotically correct as $\xi \to \infty$. This assumption is usually justified because the structural design aims at reaching very small failure probabilities. In the case of independence, the probabilities can simply be multiplied

$$1 - P_E(\xi, T) = \mathbf{Prob}[a_1 < \xi] \cdot \mathbf{Prob}[a_2 < \xi] \cdot \ldots \cdot \mathbf{Prob}[a_N < \xi] \tag{4.106}$$

If the response process $X(t)$ is stationary then the individual event probabilities are identical and we get

$$1 - P_E(\xi, T) = \mathbf{Prob}[a_i < \xi]^N \tag{4.107}$$

Finally, we can compute the first passage probability from this

$$P_E(\xi, T) = 1 - \mathbf{Prob}[a_i < \xi]^N \tag{4.108}$$

For the remaining computation of the exceedance probability of a single maximum a_i we assume that the process $X(t)$ is Gaussian with a mean value $\bar{X}$ and a standard deviation σ_X. Assuming narrow-bandedness we have (Lin 1976, cf. section 4.2.1)

$$\mathbf{Prob}[a_i \geq \xi] = P_\xi = \exp\left[-\frac{(\xi - \bar{X})^2}{2\sigma_X^2}\right]; \quad \xi > \bar{X} \tag{4.109}$$

and from this

$$\mathbf{Prob}[a_i < \xi] = 1 - P_\xi = 1 - \exp\left[-\frac{(\xi - \bar{X})^2}{2\sigma_X^2}\right]; \quad \xi > \bar{X} \tag{4.110}$$

The random number N of local maxima in the time interval $[0, T]$ is replaced by its expected value $\bar{N}$. For narrow-band random processes with a central circular frequency

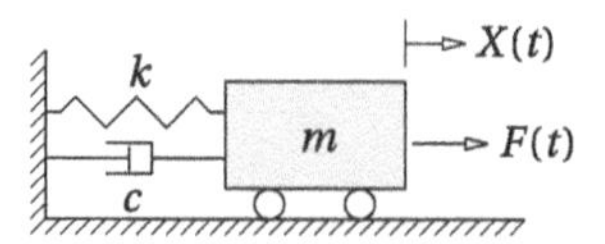

Figure 4.17 SDOF system subjected to random load $F(t)$.

ω_0 (this is typically the fundamental natural circular frequency of the structure) $\bar{N}$ is determined by

$$\bar{N} = \frac{\omega_0 T}{2\pi} \tag{4.111}$$

Example 4.4 (Design of a damper)
A single-degree-of-freedom system (cf. Fig. 4.17) is loaded by a stationary random process $F(t)$ with mean value $\bar{F} = 0.25\,\text{N}$ and constant power spectral density $S_{FF}(\omega) = 0.0003\,\text{N}^2\text{s}$. The damper c is to be designed such that the probability $P_E(\xi, T)$ of exceeding the threshold $\xi = 0.5\,\text{m}$ within a time interval of length $T = 60\,\text{s}$ is smaller than 10^{-4}. The mean value of the displacement is readily computed as $\bar{X} = \bar{F}/k = 0.25\,\text{m}$. The natural circular frequency of the system is $\omega_0 = \sqrt{k/m} = 1\,\text{rad/s}$. Hence the expected number of maxima in the time interval $[0, T]$ is $\bar{N} = 9.5 \approx 10$. From Eqs. 4.108 and 4.109 we obtain

$$P_E = 1 - (1 - P_\xi)^{10} \leq 10^{-4} \tag{4.112}$$

and from that

$$P_\xi \leq 10^{-5} \tag{4.113}$$

This leads to

$$P_\xi = \exp\left[-\frac{(0.50 - 0.25)^2}{2\sigma_X^2}\right] \leq 10^{-5} \tag{4.114}$$

and by rearranging to

$$\sigma_X^2 \leq 0.00271\,\text{m}^2 \tag{4.115}$$

If we finally consider Eq. 4.78, we obtain

$$c \geq \frac{\pi S_{FF}(\omega_0)}{k\sigma_X^2} = 0.347\,\text{Ns/m} \tag{4.116}$$

This corresponds to a damping ratio of $\zeta = 0.174$.

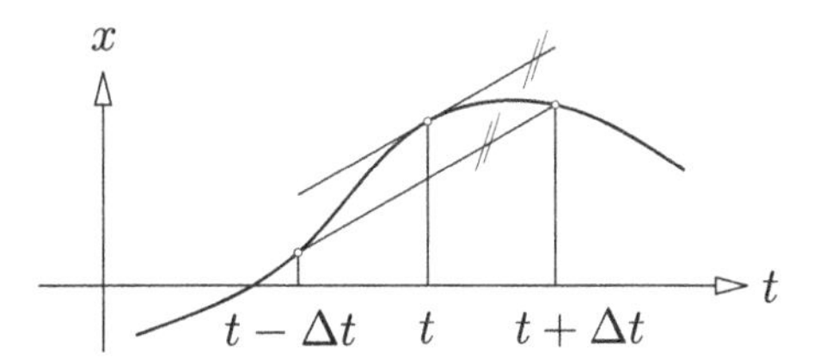

Figure 4.18 Kinematic assumption of central difference method.

4.5 Monte-Carlo simulation

4.5.1 *General remarks*

The analysis of nonlinear effects on random vibrations is very difficult. A possible analytical approach is based on linearization methods which aim at describing the mean value and covariance function of the response as good as possible. In most cases, this so-called *Equivalent Linearization* (see e.g. Roberts and Spanos 2003) cannot represent the distribution function of the response accurately. Hence in many cases, methods which are more general and also require more computational effort based on *Monte Carlo simulation* are utilized. In this approach, artificial realizations of the excitation process $F(t)$ are digitally generated and for each excitation the response $X(t)$ is computed using a numerical scheme suitable for nonlinear dynamic structural analysis. This produces an ensemble of response sample functions which can be analyzed using statistical methods with respect to mean value, variance, first passage probability, etc. The procedure follows the diagram sketched previously in Fig. 2.21.

4.5.2 *Central difference method*

Direct integration methods attempt to approximate the response $x(t)$ directly in terms of simple mathematical functions. One very popular method is the *central difference method*. Within this approach, the response $x(t)$ is approximated by a quadratic function of t within a small time interval of length $s\Delta t$. This is shown schematically in Fig. 4.18. The velocity and acceleration at time t can be expressend in terms of the displacements

$$\dot{\mathbf{x}}(t) = \frac{\mathbf{x}(t+\Delta t) - \mathbf{x}(t-\Delta t)}{2\Delta t}$$

$$\ddot{\mathbf{x}}(t) = \frac{\mathbf{x}(t+\Delta t) - 2\mathbf{x}(t) + \mathbf{x}(t-\Delta t)}{\Delta t^2} \tag{4.117}$$

Taking into account the equations of motion (4.67) at the time instant t, we obtain

$$\left(\frac{1}{\Delta t^2}\mathbf{M} + \frac{1}{2\Delta t}\mathbf{C}\right)\mathbf{x}(t+\Delta t) = \mathbf{f}(t) - \left(\mathbf{K} - \frac{2}{\Delta t^2}\mathbf{M}\right)\mathbf{x}(t) - \left(\frac{1}{\Delta t^2}\mathbf{M} + \frac{1}{2\Delta t}\mathbf{C}\right)\mathbf{x}(t-\Delta t) \tag{4.118}$$

For given values of $\mathbf{x}(t)$ and $\mathbf{x}(t-\Delta t)$ the value of $\mathbf{x}(t+\Delta t)$ can be computed by solving a system of linear equations. In order to start the procedure it is necessary to compute the value of $\mathbf{x}(-\Delta t)$. For this, the initial values $\mathbf{x}_0$ und $\mathbf{v}_0$ at time $t=0$ are utilized. From Eq. 4.117 we immediately get

$$\mathbf{x}(-\Delta t) = \mathbf{x}_0 - \mathbf{v}_0 \Delta t + \frac{\Delta t^2}{2} \ddot{\mathbf{x}}(0) \tag{4.119}$$

Here the acceleration $\ddot{\mathbf{x}}(0)$ is computed from the equations of motion (4.67) at time $t=0$:

$$\ddot{\mathbf{x}}(0) = -\mathbf{M}^{-1} \left[\mathbf{C}\mathbf{v}_0 + \mathbf{K}\mathbf{x}_0 - \mathbf{f}(0) \right] \tag{4.120}$$

The central difference method is very simple and easy to use. It is, however, only *conditionally stable*. This means that choosing a time step larger than a critical value, i.e. $\Delta t > \Delta t_{\text{crit}}$, leads to an exponential divergence of the numerical solution. It can be shown that the central difference method has a critical time step which is defined in terms of the largest natural circular frequency ω_0 of the system

$$\Delta t \leq \Delta t_{\text{crit}} = \frac{2}{\omega_0} \tag{4.121}$$

This can be proved by reformulating the solution of the free vibration of an oscillator in terms of a recursive equation

$$\begin{bmatrix} x(t+2\Delta t) \\ x(t+\Delta t) \end{bmatrix} = \begin{bmatrix} 2 - \frac{k}{m}\Delta t^2 & -1 \\ 1 & 0 \end{bmatrix} \begin{bmatrix} x(t+\Delta t) \\ x(t) \end{bmatrix} = \mathbf{Z} \begin{bmatrix} x(t+\Delta t) \\ x(t) \end{bmatrix} \tag{4.122}$$

For the computation of multiple time steps this scheme is applied repeatedly. As a consequence, the matrix $\mathbf{Z}$ is repeatedly multiplied by itself. That leads to exponential growth of the solution if at least one eigenvalue $\lambda_{\max}$ of $\mathbf{Z}$ has a magnitude $|\lambda_{\max}| > 1$. The eigenvalues $\lambda_{1,2}$ are given by

$$\lambda_{1,2} = 1 - \frac{a}{2} \pm \sqrt{-a + \frac{a^2}{4}}; \quad a = \frac{k}{m}\Delta t^2 \tag{4.123}$$

In the case $a < 4$ the eigenvalues are complex and we have

$$|\lambda_{1,2}|^2 = \left(1 - \frac{a}{2}\right)^2 + a - \frac{a^2}{4} = 1 \tag{4.124}$$

In the case $a \geq 4$ the eigenvalues are real. Then λ_2 is the one with the larger magnitude and from

$$|\lambda_2| = \left| 1 - \frac{a}{2} - \sqrt{-a + \frac{a^2}{4}} \right| = \frac{a}{2} + \sqrt{-a + \frac{a^2}{4}} - 1 > 1 \tag{4.125}$$

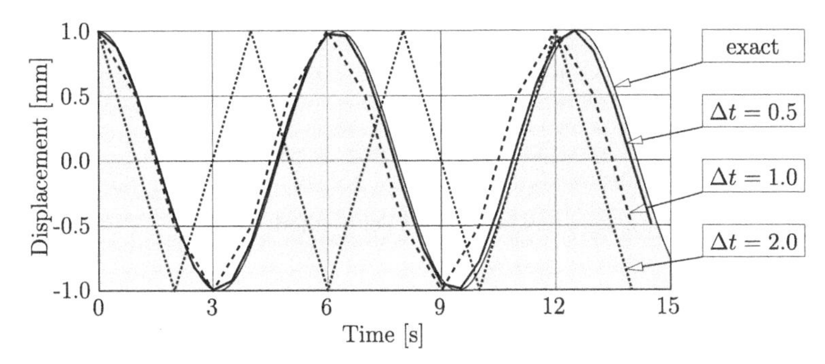

Figure 4.19 Central difference solution, system data $m = 1$ kg, $k = 1$ N/m, $c = 0$ and initial conditions $x_0 = 1$ mm, $v_0 = 0$.

we derive the condition for instability

$$\frac{a}{2} + \sqrt{-a + \frac{a^2}{4}} > 2 \rightarrow a > 4 \qquad (4.126)$$

Hence the results from this method diverge exponentially if $a > 4$ or $\Delta t > 2\sqrt{\frac{m}{k}} = \frac{2}{\omega_0}$.

Example 4.5 (Free vibration of an undamped oscillator – Central difference method)
For a single-degree-of-freedom undamped oscillator the approximate solution according to the central difference method is computed. The results are shown in Fig. 4.19 for different values of the time step Δt and compared to the exact solution. It can be seen that the amplitude of the vibration is computed correctly for all values of the time step. The frequency of the oscillation is overestimated with increasing time step size.

4.5.3 *Euler method*

This simplest method for the integration of differential equations is based on a linear approximation of the solutions over a time interval of length Δt. Since this method is suitable for systems of first-order differential equations only, structural vibration problems are recast into the phase space of displacements and velocities.

The displacement and velocity at time $t + \Delta$ are

$$\begin{aligned} \mathbf{x}(t + \Delta t) &= \mathbf{x}(t) + \dot{\mathbf{x}}(t)\Delta t \\ \dot{\mathbf{x}}(t + \Delta t) &= \dot{\mathbf{x}}(t) + \ddot{\mathbf{x}}(t)\Delta t \end{aligned} \qquad (4.127)$$

The acceleration $\ddot{\mathbf{x}}(t)$ is computed from the equation of motion (4.67) at time t.

Example 4.6 (Free vibration of an undamped oscillator – Euler method)
For a single-degree-of-freedom undamped oscillator the approximate solution according to the Euler method is computed. The results are shown in Fig. 4.21 for different values of the time step Δt and compared to the exact solution. It can be seen that the

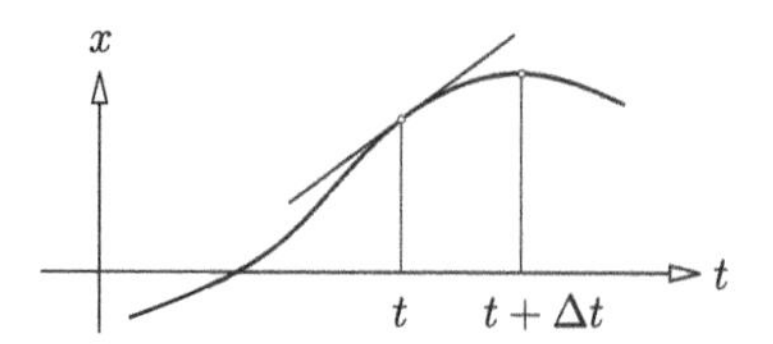

Figure 4.20 Kinematic assumption of explicit Euler method.

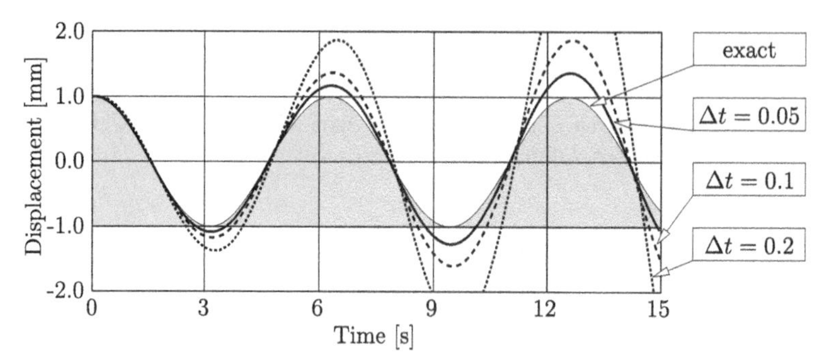

Figure 4.21 Explicit Euler solution, system data $m = 1$ kg, $k = 1$ N/m, $c = 0$ and initial conditions $x_0 = 1$ mm, $v_0 = 0$.

errors are significantly larger compared to the central difference solutions. Also, there is a continuous increase of the total energy of the system.

4.5.4 Newmark method

As an alternative to explicit methods discussed in sections 4.5.2 und 4.5.3 there are also implicit methods. One specific example is Newmark's method (Bathe 1996). The basic assumptions regarding the kinematics are comparable to the central difference method

$$\dot{\mathbf{x}}(t + \Delta t) = \dot{\mathbf{x}}(t) + \frac{\Delta t}{2}[\ddot{\mathbf{x}}(t) + \ddot{\mathbf{x}}(t + \Delta t)] \tag{4.128}$$

$$\mathbf{x}(t + \Delta t) = \mathbf{x}(t) + \dot{\mathbf{x}}(t)\Delta t + \frac{\Delta t^2}{4}[\ddot{\mathbf{x}}(t) + \ddot{\mathbf{x}}(t + \Delta t)] \tag{4.129}$$

In addition, Eq. 4.67 is applied for the future time $t + \Delta t$

$$\mathbf{M}\ddot{\mathbf{x}}(t + \Delta t) + \mathbf{C}\dot{\mathbf{x}}(t + \Delta t) + \mathbf{K}\mathbf{x}(t + \Delta t) = \mathbf{f}(t + \Delta t) \tag{4.130}$$

Together, this leads to

$$\left[\mathbf{K} + \frac{4}{\Delta t^2}\mathbf{M} + \frac{2}{\Delta t}\mathbf{C}\right]\mathbf{x}(t + \Delta t) = \mathbf{f}(t + \Delta t) + \mathbf{M}\left[\frac{4}{\Delta t^2}\mathbf{x}(t) + \frac{4}{\Delta t}\dot{\mathbf{x}}(t) + \ddot{\mathbf{x}}(t)\right] + \mathbf{C}\left[\frac{2}{\Delta t}\mathbf{x}(t) + \dot{\mathbf{x}}(t)\right] \tag{4.131}$$

This procedure is unconditionally stable, i.e. it produces numerically stable results for arbitrarily large time steps Δt. If the time step is chosen to be very large, the method yields a static solution which can be seen by inspecting Eq. 4.131. The unconditional stability of Newmark's method can be shown by investigating the free vibration of an undamped oscillator with mass m and stiffness k and writing the solution procedure in terms of a recursion formula

$$\begin{bmatrix} x(t+\Delta t) \\ \dot{x}(t+\Delta t) \end{bmatrix} = \frac{1}{4m+\Delta t^2} \begin{bmatrix} 4m - k\Delta t^2 & 4m\Delta t \\ -4k\Delta t & 4m - k\Delta t^2 \end{bmatrix} \begin{bmatrix} x(t) \\ \dot{x}(t) \end{bmatrix} = \mathbf{Z} \begin{bmatrix} x(t) \\ \dot{x}(t) \end{bmatrix} \tag{4.132}$$

The eigenvalues of the matrix $\mathbf{Z}$ are

$$\lambda_{1,2} = \frac{1}{4m+k\Delta t^2}(4m - k\Delta t^2 \pm \mathrm{i}4\sqrt{km}\Delta t) \tag{4.133}$$

The absolute values of these eigenvalues are

$$\begin{aligned} |\lambda_{1,2}| &= \frac{1}{4m+k\Delta t^2}\sqrt{(4m-k\Delta t^2)^2 + 16km\Delta t^2} \\ &= \frac{1}{4m+k\Delta t^2}\sqrt{\left(4m+k\Delta t^2\right)^2} = 1 \end{aligned} \tag{4.134}$$

independent of the time interval Δt. The mathematical operations required to carry out this proof are given in the maxima-code shown in Listing 4.7.

```
1  assume(m>0,k>0);
2  s:solve([(k+4*m/dt^2)*x1=4*m/dt^2*x0+4*m/dt*y0-k*x0,
       y1=y0-dt/2*k/m*x0 - dt/2*k/m*x1],[x1,y1]);
3  xx1:rhs(s[1][1]);
4  z11:coeff(expand(xx1),x0);
5  z12:coeff(expand(xx1),y0);
6  xx2:rhs(s[1][2]);
7  z21:coeff(expand(xx2),x0);
8  z22:coeff(expand(xx2),y0);
9  z:matrix([z11,z12],[z21,z22]);
10 e:eigenvalues(z);
11 a:abs(e);
12 tex(%);
13 ratsimp(%);
```

Listing 4.7 Proof of unconditional stability of Newmark's method.

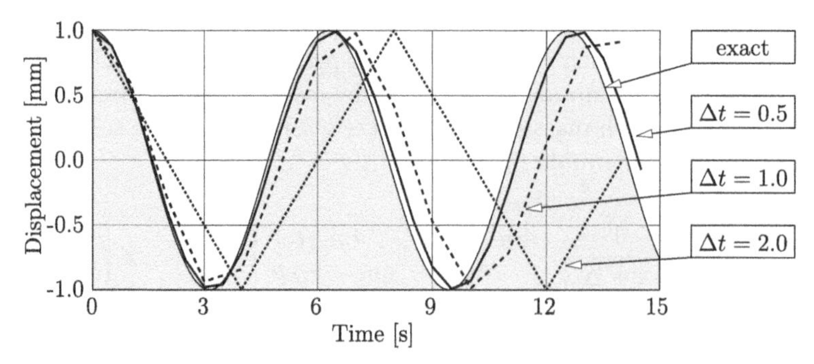

Figure 4.22 Implicit Newmark solution, system data $m = 1$ kg, $k = 1$ N/m, $c = 0$ and initial conditions $x_0 = 1$ mm, $v_0 = 0$.

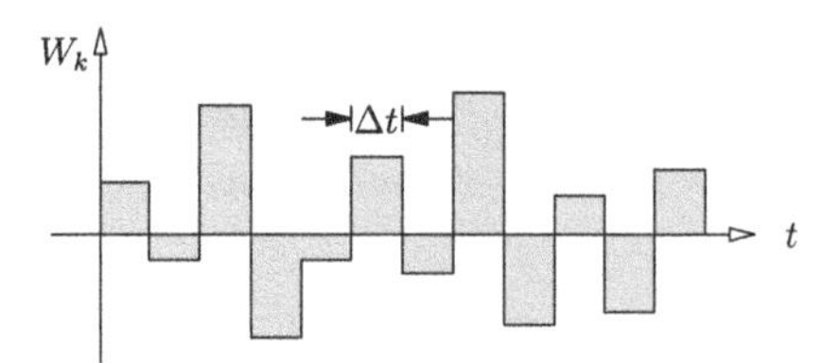

Figure 4.23 Discrete representation of white noise.

Example 4.7 (Free vibration of an undamped oscillator – Newmark method)
For a single-degree-of-freedom undamped oscillator the approximate solutions according to Newmark's method is computed. The results are shown in Fig. 4.22 for different values of the time step Δt and compared to the exact solution. It can be seen that the errors are rather small and that the frequency of the oscillator is underestimated.

4.5.5 *Digital simulation of white noise*

In order to apply this approach in digital simulation, the continuous time white noise excitation driving the Kanai-Tajimi filter needs to be discretized. This is achieved by representing the white noise $w(t)$ by a sequence of i.i.d. random variables W_k assumed to have constant values spaced at time intervals Δt (cf. Fig. 4.23). The variables W_k have zero mean and a variance $\sigma^2_{W_k}$ which is related to the intensity D_0 of the white noise (or its spectral density S_0) and the time interval Δt by

$$\sigma^2_{W_k} = \frac{D_0}{\Delta t} = \frac{2\pi S_0}{\Delta t} \tag{4.135}$$

For purposes of the subsequent reliability analysis, the white noise is conveniently represented by a sequence of i.i.d. random variables U_k with unit standard deviation. The variables W_k are then generated by

$$W_k = \sqrt{\frac{D_0}{\Delta t}} U_k = \sqrt{\frac{2\pi S_0}{\Delta t}} U_k \tag{4.136}$$

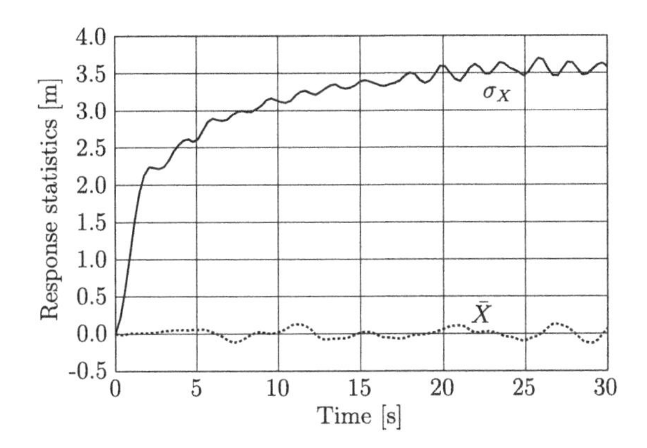

Figure 4.24 Statistics of response of Duffing oscillator to white noise excitation (1000 samples).

Example 4.8 (Duffing oscillator)
As an example, consider a well-studied simple nonlinear oscillator defined by the equation of motion

$$m\ddot{X} + c\dot{X} + k(X + \alpha X^3) = F(t) \tag{4.137}$$

The system parameters k and m must be positive and c and α must be non-negative. We assume that the excitation $F(t)$ is zero-mean white noise with spectral density S_0. Solving the equation of motion numerically using the central difference method we can obtain statistics of the response $X(t)$. The numerical values for system and excitation parameters are $k=1$, $m=1$, $c=0.1$, $\alpha=0.05$ and $S_0=1$.

The procedure leading to the results as shown in Fig. 4.24 is described in the `octave`-code below

```
1  NSIM=1000;
2  tu = 0;
3  to = 30;
4  NT = 100;
5  dt = (to-tu)/NT;
6  t = linspace(tu,to,NT+1);
7  S0=1
8
9  m = 1;
10 k = 1;
11 c = 0.1;
12 alpha = 0.05;
13 sim=[];
14
15 for i=1:NSIM
16   x=0;
```

```
17     x0 = 0;
18     f = randn()*sqrt(2*pi*S0/dt);
19     xm1 = f/m*dt^2/2;
20     for j=1:NT
21       a = (f+c*xm1/2/dt-k*(x0+alpha*x0^3))/m;
22       x1 = a*dt^2 + 2*x0 - xm1;
23       x1 = x1/(1+c/2/m*dt);
24       f = randn()*sqrt(2*pi*S0/dt);
25       xm1 = x0;
26       x0 = x1;
27       x=[x,x0];
28     end
29     sim=[sim;x];
30   end
31   m = mean(sim);
32   s = std(sim);
33   plot(t,m,t,s);
34   pause;
35   res=[t',m',s'];
36   save('-ascii', 'duffstat.txt', 'res');
```

It can be seen that the mean value $\bar{X}$ fluctuates about zero whereas the standard deviation σ_X after initial increase from zero approaches a stationary value of about 3.5.

Example 4.9 (Elasto-plastic system)
As the next example, consider a simple elasto-plastic oscillator as shown in Fig. 4.25. The elasto-plastic behavior is modeled by two springs k_1 and k_2 and a friction element with a limiting friction force s. This friction element is activated (and with it the plastic displacement z) if the force in the spring k_2 exceeds the friction limit s. The total restoring force r of the spring-friction assembly is given by

$$r = k_1 x + k_2(x - z) \tag{4.138}$$

The effect of plasticity is described by the rate of change of z. This rate can be expressed as

$$\dot{z} = h(x, \dot{x}, z)\dot{x} \tag{4.139}$$

Here, several cases must be considered for the function h, i.e.

$$h = \begin{cases} 0 & |k_2(x - z)| < s \\ 0 & k_2(x - z) > s \wedge \dot{x} < 0 \\ 0 & k_2(x - z) < -s \wedge \dot{x} > 0 \\ 1 & \text{else} \end{cases} \tag{4.140}$$

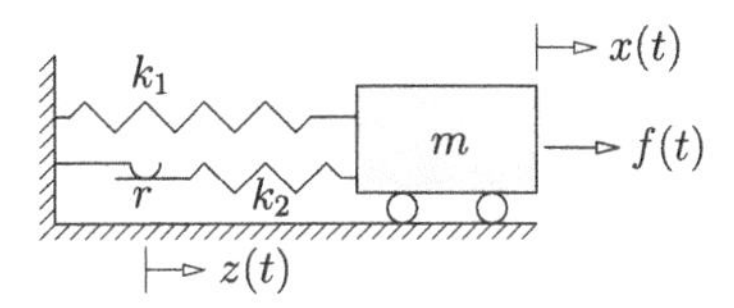

Figure 4.25 Elasto-plastic oscillator.

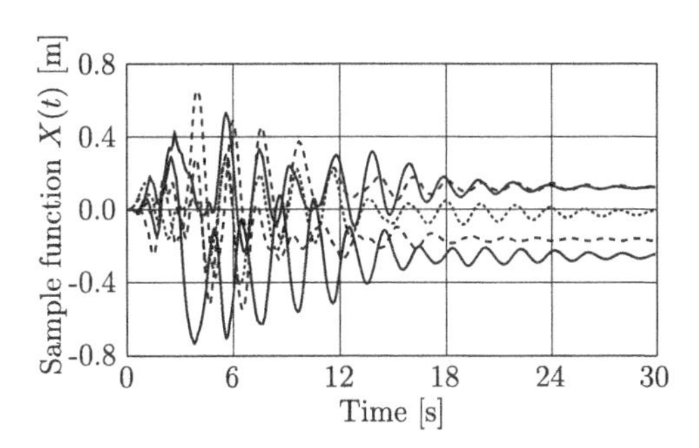

Figure 4.26 Sample functions of displacement response X.

The tangential stiffness k_T of the systems depends on h

$$k_T = k_1 + [1 - h(x, \dot{x}, z)]k_2 \tag{4.141}$$

The excitation $F(t)$ is assumed to be amplitude-modulated white noise, i.e.

$$F(t) = e(t) \cdot W(t) \tag{4.142}$$

in which

$$e(t) = 4 \cdot [\exp(-0.25t) - \exp(-0.5t)] \tag{4.143}$$

and $W(t)$ is a white noise with spectral density $S_0 = \frac{1}{2\pi}$. Here, the equations of motion are reformulated in first-order form which facilitates the treatment of the internal plastic variable z.

Fig. 4.26 shows some generated sample functions of the displacement response $X(t)$. The results for the mean value and standard deviation of the displacement X and the internal plastic displacement Z are shown in Figs. 4.27 and 4.28. The results were obtained by averaging over 1000 samples and therefore are affected by statistical uncertainty. This can easily be seen by considering the mean value functions $\bar{X}(t)$ and $\bar{Z}(t)$ which theoretically should be zero. The octave-code generating these samples and carrying out the statistical analysis is given in Listing 4.8. This computation is based on the explicit Euler time integration scheme.

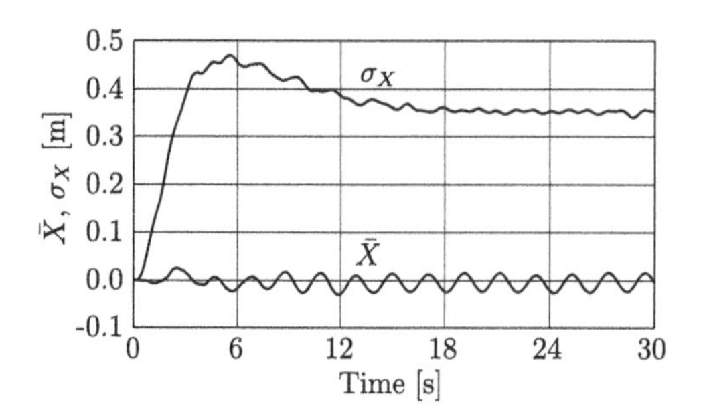

Figure 4.27 Mean value and standard deviation of displacement X.

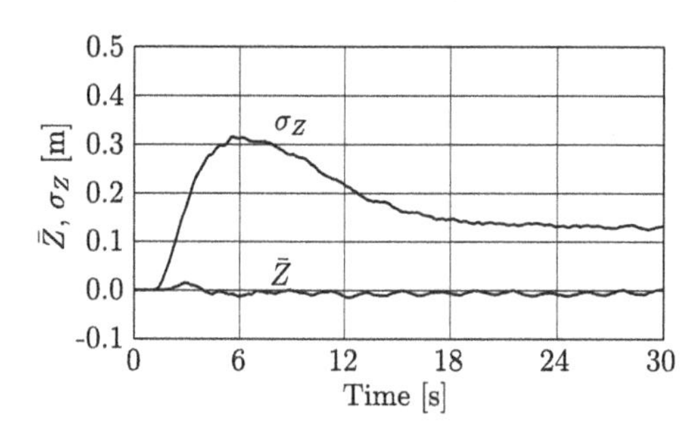

Figure 4.28 Mean value and standard deviation of plastic displacement Z.

```
1  function env=env(t)
2  env=4*(exp(-t/4)-exp(-t/2));
3  endfunction
4
5  function ydot=ydot(y,f)
6  global k1 k2 c;
7  x=y(1);
8  v=y(2);
9  z=y(3);
10 s=3;
11 fr=k2*(x-z);
12 ft=k1*x+fr;
13 ft+=c*v;
14 vd=f-ft;
15 xd = v;
16 if (abs(fr)<s || (fr>s && v<0) || (fr<-s && v>0))
17   zd=0;
18 else
19   zd = v;
20 endif
21 ydot(1) = xd;
22 ydot(2) = vd;
```

Listing 4.8 Compute Monte Carlo samples for oscillator with dry friction.

```
23  ydot(3) = zd;
24  endfunction
25
26  dt = .1;
27  T=30;
28  N=T/dt;
29  M=1000;
30  t=0; force=0;
31  global k1 k2 c;
32  k1=1;
33  k2=9;
34  c=.4;
35
36  x1=zeros(M,N+1);
37  x2=x1; x3=x1;
38  ff=x1;
39  tt = linspace(0,N*dt,N+1);
40  for j=1:M
41    x=[0,0,0];
42    j
43    for i=1:N
44      t = tt(i);
45      force = randn()*env(t)/sqrt(dt);
46      xd = ydot(x,force);
47      x(1) += xd(1)*dt;
48      x(2) += xd(2)*dt;
49      x(3) += xd(3)*dt;
50      x1(j,i+1) = x(1);
51      x2(j,i+1) = x(2);
52      x3(j,i+1) = x(3);
53      ff(i) = force;
54    endfor
55  endfor
56  m1 = mean(x1);
57  s1 = std(x1);
58  m3 = mean(x3);
59  s3 = std(x3);
60
61  fid = fopen(" mc_plast_stat.txt","w");
62  for i=1:N+1
63    fprintf(fid, "%g %g %g %g %g\n", tt(i),m1(i),s1(i),
        m3(i),s3(i));
64  endfor
65  fclose(fid);
66  plot(tt,m1,tt,s1,tt,m3,tt,s3);
67  pause
```

Listing 4.8 Continued

4.6 Fokker-Planck equation

For systems under white noise excitation (stationary or non-stationary), the equations of motion (4.67) can be written in Itô form (Itô 2004; Lin 1976; Lin and Cai 1995)

$$\mathrm{d}\mathbf{Y} = \boldsymbol{\mu}\,\mathrm{d}t + \boldsymbol{\sigma}\mathrm{d}B(t) \tag{4.144}$$

Here, $\mathbf{Y} = [Y_1, \ldots, Y_n]^T$ is the random state vector, $\boldsymbol{\mu}$ is the *drift*, and $\boldsymbol{\sigma}$ is the *diffusion*. $B(t)$ denotes a process with independent increments, a so-called *Wiener process*. The second-order statistical properties of $B(t)$ are given by

$$\mathbf{E}[B(t)] = 0; \ \mathbf{E}[B(t_1)B(t_2)] = \min(t_1, t_2) \tag{4.145}$$

As a consequence, the statistical properties of the increments $\mathrm{d}B(t) = B(t+\mathrm{d}t) - B(t)$ are

$$\mathbf{E}[\mathrm{d}B(t)] = 0; \ \mathbf{E}[\mathrm{d}B(t)^2] = \mathrm{d}t \tag{4.146}$$

The latter property indicates that $\mathrm{d}B(t)$ grows with $\sqrt{\mathrm{d}t}$ in the mean square sense. This implies that standard differential calculus cannot be applied, since this requires growth with $\mathrm{d}t$. In order to derive the Itô equation for functions $g(\mathbf{Y})$ of a random process $\mathbf{Y}$ governed by an Itô equation such as Eq. 4.144, more than the first-order terms must be considered in the Taylor expansion.

The solution of an Itô equation forms a Markov vector process, which implies that the joint probability density function of its components is described by an initial density function at time t_0 and the transition probability density function defining the conditional probability density function at time t given the values at time t_0.

$$f_{\mathbf{Y}}(\mathbf{y}, t) = f_{\mathbf{Y}}(\mathbf{z}, t_0) p_{\mathbf{Y}}(\mathbf{y}, t|\mathbf{z}, t_0) \tag{4.147}$$

The transition probability density function is governed by the Fokker-Planck equation, a partial differential equation

$$\frac{\partial p}{\partial t} + \sum_{i=1}^{n} \frac{\partial}{\partial y_i}(\mu_i p) - \frac{1}{2}\sum_{i=1}^{n}\sum_{k=1}^{n} \frac{\partial^2}{\partial y_i \partial y_k}\left(\sum_{\ell=1}^{n} \sigma_{i\ell}\sigma_{k\ell}\, p\right) = 0 \tag{4.148}$$

Consider the special case that a SDOF-oscillator is governed by the differential equation

$$m\ddot{X} + h(E)\dot{X} + g(X) = W(t) \tag{4.149}$$

in which

$$E(X,\dot{X}) = \frac{1}{2}m\dot{X}^2 + \int_0^X g(\xi)d\xi \tag{4.150}$$

denotes the sum of kinetic and deformation energy of the oscillator and $W(t)$ is white noise with spectral density S_0. The state vector in this case is

$$\mathbf{Y} = \begin{bmatrix} X \\ \dot{X} \end{bmatrix} \tag{4.151}$$

and the drift and diffusion terms become

$$\boldsymbol{\mu} = \begin{bmatrix} -\dot{X} \\ -\frac{1}{m}g(X) - \frac{h(E)}{m}\dot{X} \end{bmatrix}; \quad \boldsymbol{\sigma} = \begin{bmatrix} 0 \\ \frac{1}{m}\sqrt{2\pi S_0} \end{bmatrix} \tag{4.152}$$

For this case, the Fokker-Planck equation becomes

$$\frac{\partial p}{\partial t} + \dot{x}\frac{\partial p}{\partial x} + \frac{\partial}{\partial \dot{x}}\left[\left(-\frac{g(x)}{m} - \frac{h(E)\dot{x}}{m}\right)p\right] - \frac{\pi S_0}{m^2}\frac{\partial^2 p}{\partial \dot{x}^2} = 0 \tag{4.153}$$

For the so-called stationary state (which occurs when $t \to \infty$) there exists a closed-form solution of the Fokker-Planck-equation (Lin 1976)

$$f_{X,\dot{X}} = C\exp\left(-\frac{1}{\pi S_0}\int_0^E h(\eta)d\eta\right) \tag{4.154}$$

Here, C is a normalizing constant. Other special solutions involving both external and parametric random excitations are given in Dimentberg 1982.

Example 4.10 (Stationary probability density of Duffing oscillator)
For the Duffing oscillator as described by Eq. 4.137 the total energy is given by

$$E = \frac{1}{2}m\dot{X}^2 + k\left(\frac{X^2}{2} + \alpha\frac{X^4}{4}\right) \tag{4.155}$$

Since the damping force does not depend on the energy, we simply have $H(E) = c$. Therefore the joint probability density function of the response and its time derivative is given by

$$f_{X,\dot{X}}(x,\dot{x}) = C\exp\left(-\frac{c}{\pi S_0}E\right) = C\exp\left(-\frac{c}{\pi S_0}\left[k\left(\frac{x^2}{2} + \alpha\frac{x^4}{4}\right) + \frac{1}{2}m\dot{x}^2\right]\right) \tag{4.156}$$

Obviously, this can be separated into a product of a function of x and another function of $\dot{x}$. Hence the response X and its time derivative $\dot{X}$ are independent (rather than being merely uncorrelated) when evaluated at the same time. Also, it can be seen that the marginal probability density function of the velocity is Gaussian:

$$f_{\dot{X}} = C_2 \exp\left(-\frac{cm}{\pi S_0}\frac{\dot{x}^2}{2}\right) \tag{4.157}$$

From this it can be seen that (compare this to Eq. 4.78)

$$\sigma_{\dot{X}}^2 = \frac{\pi S_0}{cm} \tag{4.158}$$

and therefore

$$C_2 = \frac{1}{\sqrt{2\pi}\sigma_{\dot{X}}} \tag{4.159}$$

The probability density function is then given by

$$f_X(x) = C_1 \exp\left[-\frac{kc}{\pi S_0}\left(\frac{x^2}{2} + \alpha\frac{x^4}{4}\right)\right] \tag{4.160}$$

in which C_1 has to be determined by integration

$$\frac{1}{C_1} = \int_{-\infty}^{\infty} \exp\left[-\frac{kc}{\pi S_0}\left(\frac{x^2}{2} + \alpha\frac{x^4}{4}\right)\right] \mathrm{d}x \tag{4.161}$$

The variance can then be obtained by integration as well

$$\sigma_X^2 = C_1 \int_{-\infty}^{\infty} x^2 \exp\left[-\frac{kc}{\pi S_0}\left(\frac{x^2}{2} + \alpha\frac{x^4}{4}\right)\right] \mathrm{d}x \tag{4.162}$$

For the numerical values $k = 1$, $m = 1$, $c = 0.1$, $\alpha = 0.05$ and $S_0 = 1$ we obtain a standard deviation $\sigma_X = 3.54$ m. This compares quite well to the Monte Carlo result as shown in Fig. 4.24. The values of the kurtosis which is an indicator of the deviation from a Gaussian distribution is $\kappa = -0.61$. For an increased value of $\alpha = 0.2$ the standard deviation drops to 2.69 and the kurtosis becomes -0.71. The probability density functions $f_X(x)$ are shown in Fig. 4.29 for both values of α. The `octave`-code for these computations is given in Listing 4.9.

4.7 Statistical linearization

4.7.1 *General concept*

Due to the complexity of nonlinear system analysis even in the deterministic case, it has frequently been suggested to replace the actual nonlinear system by an equivalent linear system. Of course, one of the major problems lies in the appropriate definition of equivalence. In stochastic analysis, a first goal for the analysis is the computation of second-order statistics. This means, an equivalent system for this purpose should

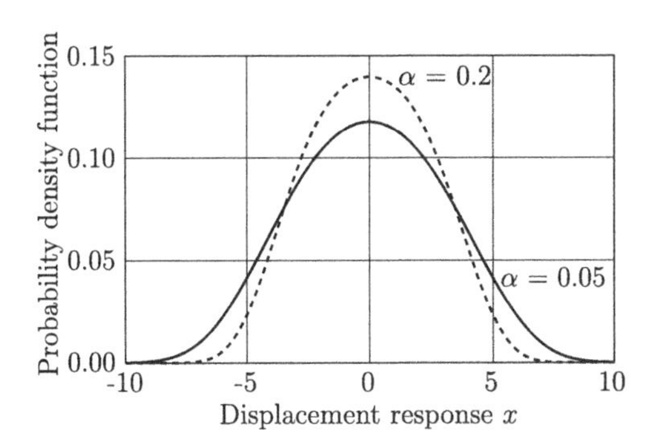

Figure 4.29 Probability density function of response of Duffing oscillator.

```
1  k=1;
2  m=1;
3  c=0.1;
4  alfa = 0.05;
5  S0=1;
6
7  xu=-30;
8  xo=30;
9  NX=401;
10 x=linspace(xu,xo,NX);
11 dx=(xo-xu)/NX;
12 f=k*(x.^2/2+alfa*x.^4/4);
13 f=f*c/pi/S0;
14 f=exp(-f);
15 C1=sum(f)*dx;
16 var=sum(x.^2.*f/C1)*dx;
17 sig=sqrt(var)
18 kurt=sum(x.^4.*f/C1)*dx/var^2 - 3
19 pdf = [x',f'/C1];
20 save('-ascii', 'dufpdf05.txt', 'pdf');
```

Listing 4.9 Normalize probability density function for Duffing oscillator and compute standard deviation and kurtosis.

reproduce the correct mean value and covariance functions of the system response. The statistical linearization method is extensively treated in Roberts and Spanos 2003, and the subsequent representation follows this reference.

Assume that the system under consideration has a governing equation of motion in the form of Eq. 4.68. In an equivalent linear system, the function **g** is replaced by the linear function

$$\hat{\mathbf{g}}(\mathbf{x}, \dot{\mathbf{x}}) = \mathbf{C}_e\dot{\mathbf{x}} + \mathbf{K}_e\mathbf{x} \tag{4.163}$$

Here, $\mathbf{C}_e$ denotes the equivalent damping matrix and $\mathbf{K}_e$ is the equivalent stiffness matrix. The objective is to determine $\mathbf{C}_e$ and $\mathbf{K}_e$ such that the response of the equivalent linear system matches the response of the nonlinear system in a second-moment sense. Since the difference between the two systems lies in the representation of $\mathbf{g}$ by $\hat{\mathbf{g}}$, it is fairly natural to introduce an error measure $\boldsymbol{\epsilon}$ by

$$\boldsymbol{\epsilon} = \mathbf{g} - \hat{\mathbf{g}} = \mathbf{g}(\mathbf{x}, \dot{\mathbf{x}}) - \mathbf{C}_e\dot{\mathbf{x}} - \mathbf{K}_e\mathbf{x} \tag{4.164}$$

and to attempt to make this error small in an appropriate way. For stochastic analysis it is useful to minimize the expected value of the squared Euclidian norm of $\boldsymbol{\epsilon}$:

$$E = \mathbf{E}[\boldsymbol{\epsilon}^T\boldsymbol{\epsilon}] \rightarrow \text{Min.!} \tag{4.165}$$

When inserting (4.164) into this equation, we obtain

$$\begin{aligned} E &= \mathbf{E}\left[(\mathbf{g} - \hat{\mathbf{g}})^T(\mathbf{g} - \hat{\mathbf{g}})\right] \\ &= \mathbf{E}\left[(\mathbf{g} - \mathbf{C}_e\dot{\mathbf{x}} - \mathbf{K}_e\mathbf{x})^T(\mathbf{g} - \mathbf{C}_e\dot{\mathbf{x}} - \mathbf{K}_e\mathbf{x})\right] \\ &= \mathbf{E}\left[\mathbf{g}^T\mathbf{g} + \dot{\mathbf{x}}^T\mathbf{C}_e^T\mathbf{C}_e\dot{\mathbf{x}} + \mathbf{x}^T\mathbf{K}_e^T\mathbf{K}_e\mathbf{x} - 2\mathbf{g}^T\mathbf{C}_e\dot{\mathbf{x}} - 2\mathbf{g}^T\mathbf{K}_e\mathbf{x} + 2\dot{\mathbf{x}}^T\mathbf{C}_e^T\mathbf{K}_e\mathbf{x}\right] \end{aligned} \tag{4.166}$$

Formally the elements C_{ik} and K_{ik} of the equivalent damping and stiffness matrices can be obtained by taking partial derivatives and setting these derivatives equal to zero. Practically, however, this poses a major difficulty since the expected values required to solve the resulting equations are usually not known. In fact, the whole purpose of the linearization procedure is to compute some of these expectations.

It is therefore helpful to make certain assumptions on the probability density function of the response components. The assumption of jointly Gaussian distributed response vector components is made frequently. This reduces the problem of computing expected values to the mean vector and the covariance matrix.

In order to illustrate the difficulties, we will discuss a SDOF system with a cubic nonlinearity, the so-called Duffing oscillator.

Example 4.11 (Linearization of the Duffing oscillator)
Consider again a single-degree-of-freedom system with the equation of motion given in (4.137). This system has linear damping, hence it is reasonable to assume the equivalent linear system has the same damping. This means that the nonlinear function $g(X) = k(X + \alpha X^3)$ is to be replaced by the equivalent linear function

$$\hat{g}(X) = k_e X \tag{4.167}$$

The error term to be minimized then becomes

$$\begin{aligned} E &= \mathbf{E}[(kX + k\alpha X^3 - k_e X)^2] \\ &= \mathbf{E}[k^2X^2 + k^2\alpha^2X^6 + k_e^2X^2 + 2k^2\alpha X^4 - 2kk_eX^2 - 2kk_e\alpha X^4] \\ &= k^2\alpha^2\mathbf{E}[X^6] + 2k(k - k_e)\alpha\mathbf{E}[X^4] + (k - k_e)^2\mathbf{E}[X^2] \end{aligned} \tag{4.168}$$

Taking the derivative w.r.t. k_e we get

$$-2\alpha k\mathbf{E}[X^4] - 2(k - k_e)\mathbf{E}[X^2] = 0 \tag{4.169}$$

resulting in

$$k_e = k\left(1 + \alpha\frac{\mathbf{E}[X^4]}{\mathbf{E}[X^2]}\right) \tag{4.170}$$

In order to use this result, the second and fourth moments of the response must be known at each point in time.

Assuming that the mean value of the load $\bar{F}$ is zero, it can be shown that the mean value of the response $\bar{X}$ is zero as well. We can then assume the probability density function

$$f_X(x) = \frac{1}{\sqrt{2\pi}\sigma_X}\exp\left[-\frac{x^2}{2\sigma_X^2}\right] \tag{4.171}$$

in which the variance σ_X^2 is not yet known. Based on this assumption we get (cf. Eq. 2.33)

$$\mathbf{E}[X^2] = \sigma_X^2; \quad \mathbf{E}[X^4] = 3\sigma_X^4 \tag{4.172}$$

With this result we can solve for the equivalent linear stiffness in terms of the response variance

$$k_e = k(1 + 3\alpha\sigma_X^2) \tag{4.173}$$

If in addition we assume F to be white noise with power spectral density S_0 (cf. Eq. 4.58), then the variance of the response of the equivalent linear system can be computed according to Eq. 4.78

$$\sigma_X^2 = \frac{\pi S_0}{ck_e} = \frac{\pi S_0}{ck(1 + 3\alpha\sigma_X^2)} \tag{4.174}$$

Finally, this can be solved for the response variance

$$\sigma_X^2 = \frac{1}{6\alpha}\left(-1 + \sqrt{1 + 12\alpha\frac{\pi S_0}{ck}}\right) \tag{4.175}$$

Note that the term $\sigma_0^2 = \frac{\pi S_0}{ck}$ is the variance of the response of a linear system with stiffness k (i.e. without cubic nonlinearity). Therefore we can write the variance of the response of the equivalent linear system as

$$\sigma_X^2 = \frac{1}{6\alpha}\left(-1 + \sqrt{1 + 12\alpha\sigma_0^2}\right) \tag{4.176}$$

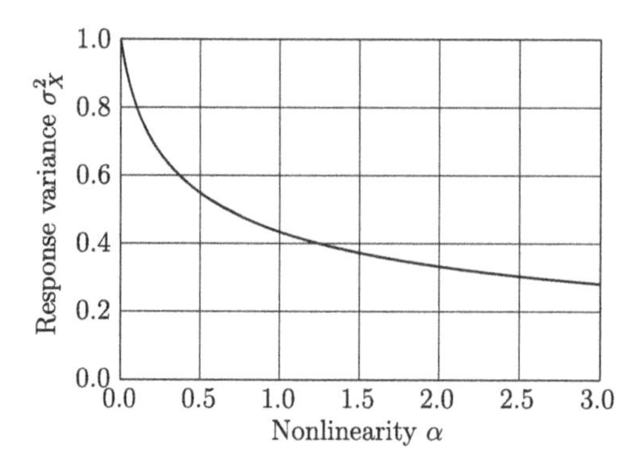

Figure 4.30 Variance of response of equivalent linear system by statistical linearization.

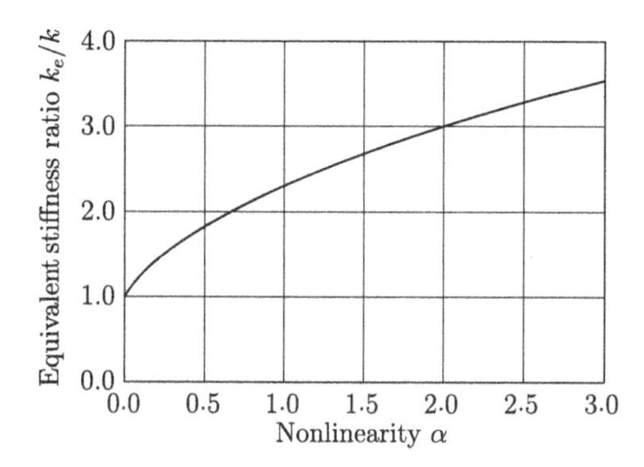

Figure 4.31 Stiffness of equivalent linear system by statistical linearization.

The dependence of σ_X^2 on the nonlinearity parameter α is shown in Fig. 4.30 for the case $\sigma_0^2 = 1$. From the variance, the equivalent stiffness k_e can be explicitly computed. The ratio $\frac{k_e}{k}$ is shown in Fig. 4.31 as a function of the nonlinearity parameter α.

For comparison, using numerical values $k = 1$, $m = 1$, $c = 0.1$, $\alpha = 0.05$ and $S_0 = 1$ we obtain a standard deviation $\sigma_X = \sqrt{\frac{1}{6 \cdot 0.05}\left(-1 + \sqrt{1 + 12 \cdot 0.05 \frac{\pi}{0.1}}\right)} = 3.39$. For the value of $\alpha = 0.20$, the result is $\sigma_X = 2.54$. Both results are reasonably close to the exact solution given in section 4.6 with errors of 4% and 6%, respectively.

Example 4.12 (Equivalent linearization of oscillator with nonlinear damping) Consider an oscillator with dry friction described by the equation of motion

$$m\ddot{X} + r\text{sign}(\dot{X}) + kX = W(t) \tag{4.177}$$

Compute the variance of the response using equivalent linearization assuming a Gaussian distribution for the response and its time derivative.

Here it is appropriate to minimize the mean square difference between the exact damping force due to dry friction and the linearized damping

$$E = \mathbf{E}[(r\text{sign}(\dot{X}) - c_e\dot{X})^2] = r^2\mathbf{E}[\text{sign}(\dot{X})^2] - 2rc_e\mathbf{E}[\text{sign}(\dot{X})\dot{X}] + c_e^2\mathbf{E}[\dot{X}^2] \quad (4.178)$$

The expectations in this equation are readily computed as

$$\begin{aligned}
\mathbf{E}[\text{sign}(\dot{X})^2] &= 1 \\
\mathbf{E}[\text{sign}(\dot{X})\dot{X}] &= 2\int_0^\infty \frac{\dot{x}}{\sqrt{2\pi}\sigma_{\dot{X}}} \exp\left(-\frac{\dot{x}^2}{2\sigma_{\dot{x}}^2}\right) \mathrm{d}\dot{x} = \sqrt{\frac{2}{\pi}}\sigma_{\dot{X}} \\
\mathbf{E}[\dot{X}^2] &= \sigma_{\dot{x}}^2
\end{aligned} \quad (4.179)$$

so that

$$E = r^2 - 2rc_e\sqrt{\frac{2}{\pi}}\sigma_{\dot{X}} + c_e^2\sigma_{\dot{X}}^2 \quad (4.180)$$

Minimizing this error term with respect to c_e results in

$$-2r\sqrt{\frac{2}{\pi}}\sigma_{\dot{X}} + 2c_e\sigma_{\dot{X}}^2 = 0 \quad (4.181)$$

which gives the equivalent viscous damping constant as

$$c_e = \sqrt{\frac{2}{\pi}}\frac{r}{\sigma_{\dot{X}}} \quad (4.182)$$

Inserting this into the expression for the variances of the displacement and velocity responses we get

$$\sigma_X^2 = \frac{\pi S_0}{c_e k} = \sqrt{\frac{2}{\pi}}\frac{\pi S_0 \sigma_{\dot{X}}}{rk}; \quad \sigma_{\dot{X}}^2 = \frac{\pi S_0}{c_e m} = \frac{k}{m}\sigma_X^2 \quad (4.183)$$

Solving this for σ_X we finally obtain

$$\sigma_X = \sqrt{\frac{2}{\pi}\frac{\pi S_0}{rk}}\sqrt{\frac{k}{m}} = \sqrt{\frac{2\pi}{km}\frac{S_0}{r}} \quad (4.184)$$

4.8 Dynamic stability analysis

4.8.1 *Basics*

Stability analysis investigates the long-term behavior of motion under the influence of perturbations, e.g. Eller 1988. For a stable motion, perturbations are insignificant, the perturbed motion stays close to the unperturbed motion. In the unstable case an infinitesimal perturbation causes a considerable change of the motion. Depending on

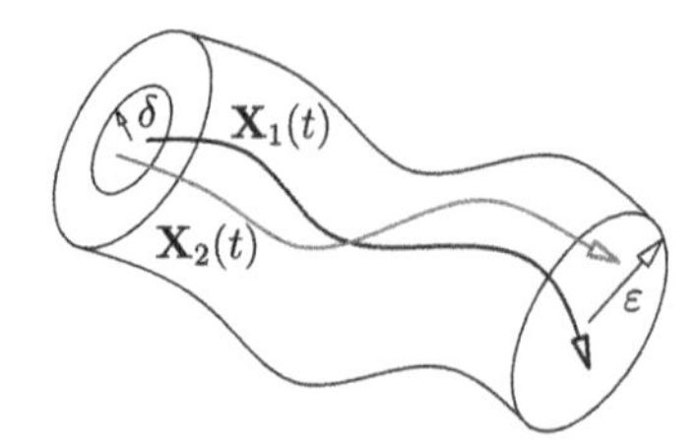

Figure 4.32 Stability in Lyapunov sense.

the type of perturbation, the stability analysis is sub-classified in structural stability and stability in Lyapunov sense. The stability concept in Lyapunov sense analyses the effect of perturbations of the initial conditions.

An unperturbed motion $\mathbf{X}_1$ is called stable in the Lyapunov sense if for any given $\epsilon > 0$ there is a $\delta(\epsilon) > 0$ so that for any perturbed motion $\mathbf{X}_2(t)$ with

$$\|\mathbf{X}_1(t_0) - \mathbf{X}_2(t_0)\| < \delta \tag{4.185}$$

we have

$$\|\mathbf{X}_1(t) - \mathbf{X}_2(t)\| < \epsilon \tag{4.186}$$

for all $t, t_0 \in \mathbb{R}^+$ (Eller 1988). In Fig.4.32 both solutions are displayed. A motion satisfying Eq. (4.186) is asymptotically stable if the condition

$$\lim_{t \to \infty} \|\mathbf{X}_1(t) - \mathbf{X}_2(t)\| = 0 \tag{4.187}$$

is fulfilled. For the stability analysis it is useful to investigate the behavior of the perturbed neighboring motion $x_2(t)$. It is only necessary to describe the long-term behavior of

$$\mathbf{Y}(t) = \mathbf{X}_2(t) - \mathbf{X}_1(t) \tag{4.188}$$

The asymptotic stability condition Eq. (4.187) gets the form

$$\lim_{t \to \infty} \|\mathbf{Y}(t)\| = 0 \tag{4.189}$$

In the case of random processes $\mathbf{X}(t)$ the stability definition as given here must be augmented by an appropriate definition of convergence. There are three generally accepted suitable definitions for convergence in a stochastic context. These definitions for extending the deterministic statement

$$\lim_{t \to t_0} \mathbf{X}(t) = \mathbf{X}_0 \tag{4.190}$$

into a stochastic context are given in the following.

Convergence in probability

$$\lim_{t \to t_0} \mathbf{Prob}[\|\mathbf{X}(t) - \mathbf{X}_0\| > \varepsilon] = 0; \ \forall \varepsilon > 0 \tag{4.191}$$

This means that the probability of getting deviations from the limit approaches zero.

Convergence in mean square

$$\lim_{t \to t_0} \mathbf{E}[\|\mathbf{X}(t) - \mathbf{X}_0\|^2] = 0 \tag{4.192}$$

This means that the variance of the difference from the limit approaches zero.

Almost sure convergence

$$\mathbf{Prob}[\lim_{t \to t_0} [\|\mathbf{X}(t) - \mathbf{X}_0\| = 0] = 1 \tag{4.193}$$

This criterion applies the limiting operation to the sample functions, and hence represents an "observable" convergence criterion. Of course, the statement that each sample converges with probability 1 does in reversal not necessarily imply that divergence is impossible. Generally, these three convergence criteria are not in agreement. One notable exception are Gaussian processes for which all three definitions give coinciding results (Lin and Cai 1995).

4.8.2 *Nonlinear stability analysis*

This section follows the presentation given by Most, Bucher, and Schorling 2004. The equation of motion of a geometrically nonlinear structural model is given by

$$\mathbf{M}\ddot{\mathbf{X}} + \mathbf{r}(\mathbf{X}, \dot{\mathbf{X}}) = \mathbf{F}(t) \tag{4.194}$$

For neighboring trajectories, the tangential equation of motion may be utilized to describe temporal evolution of the difference $\mathbf{Y}(t)$

$$\mathbf{M}\ddot{\mathbf{Y}} + \mathbf{C}\dot{\mathbf{Y}} + \mathbf{K}\mathbf{Y} = 0 \tag{4.195}$$

To analyze the dynamic stability behaviour of nonlinear systems an integration of Eq. 4.194 is necessary until stochastic stationarity is reached. In each time step, the tangential stiffness matrix $\mathbf{K}$ has to be determined. With this kind of analysis a criterion for sample stability is developed. In order to speed up explicit time integration, this equation can be projected into a subspace of dimension $m << n$ as spanned by the eigenvectors of the undamped system corresponding to the m smallest natural frequencies (Bucher 2001). These eigenvectors are the solutions to

$$(\mathbf{K}(\mathbf{X}_{stat}) - \omega_i^2 \mathbf{M})\mathbf{\Phi} = 0; \quad i = 1 \ldots m \tag{4.196}$$

In this equation, $\mathbf{X}_{stat}$ is chosen to be the displacement solution of Eq. 4.194 under static loading conditions. The mode shapes are assumed to be mass normalized.

A transformation $\mathbf{x} = \mathbf{\Phi v}$ and a multiplication of Eq.4.194 with $\mathbf{\Phi}^T$ represents a projection of the differential equation of motion for the reference solution into the subspace of dimension m as spanned by the eigenvectors

$$\ddot{\mathbf{v}} + \mathbf{\Phi}^T \mathbf{r}(\mathbf{X}, \dot{\mathbf{X}}) = \mathbf{\Phi}^T \mathbf{f} \tag{4.197}$$

The integration of Eq. 4.197 by the central difference method (Bathe 1996) requires keeping a critical time step (cf. section 4.5.2).

The time integration in the subspace and the computing of the restoring forces on the full system causes the following problem: If the start displacement or velocity vector of the time integration is not zero, for example by static loading, the projection of these vectors into the subspace is an optimiziation problem caused by the larger number of variables in the full space. By using a least square approach

$$\mathbf{v} = \mathbf{\Phi}^{-1}\mathbf{X}; \quad \mathbf{\Phi}^{-1} = \left(\mathbf{\Phi}^T\mathbf{\Phi}\right)\mathbf{\Phi}^T \tag{4.198}$$

this projection is optimally approximated but not suitable for a subspace spanned by a small number of eigenvectors. A possibility to handle this, is to start the time integration in the subspace with a displacement and velocity vector equal to zero. The start vectors have to be saved in the full system and the restoring force vector has to be computed by addition of the start and the time integration vectors (Most, Bucher, and Schorling 2004)

$$\begin{aligned} \mathbf{r}(\mathbf{X}, \dot{\mathbf{X}}) = \mathbf{r}(\mathbf{X}_{start} + \mathbf{\Phi v}, \dot{\mathbf{X}}_{start} + \mathbf{\Phi}\dot{\mathbf{v}});& \\ \mathbf{v}(t=0) = \dot{\mathbf{v}}(t=0) = 0& \end{aligned} \tag{4.199}$$

In the investigated cases the start vector $\mathbf{x}_{start}$ is the static displacement vector, the start velocities are assumed to be zero. To analyze the stability behaviour of the reference solution $\mathbf{x}_0(t)$, the long-term behavior of the neighboring motion (Eq. 4.195) is investigated. To reduce the dimension of the equation system, this equation can be projected into the same or a smaller subspace as compared to Eq. 4.197. Transformed into the state space description we obtain

$$\dot{\mathbf{z}} = \begin{bmatrix} 0 & \mathbf{I} \\ -\mathbf{\Phi}^T\mathbf{K\Phi} & -\mathbf{\Phi}^T\mathbf{C\Phi} \end{bmatrix} \mathbf{z} = \mathbf{A}[\mathbf{x}_0(t)]\mathbf{z} \tag{4.200}$$

From this equation, the Lyapunov exponent λ can be determined by a limiting process

$$\lambda(\mathbf{X}_0, \mathbf{s}) = \lim_{t\to\infty} \frac{1}{t} \log ||\mathbf{\Theta}(\mathbf{X}_0, t)\mathbf{s}|| \tag{4.201}$$

in which $\mathbf{s}$ is an arbitrary unit vector. In Eq. 4.201, $\mathbf{\Theta}(\mathbf{x}_0, t)$ is the transition matrix from time 0 to t associated with Eq. 4.200. Based on the multiplicative ergodic theorem (e.g. Arnold and Imkeller 1994) the Lyapunov exponent can also be calculated as an expected value

$$\lambda(\mathbf{X}_0, \mathbf{s}) = E\left[\frac{d}{dt} \log ||\mathbf{\Theta}(\mathbf{X}_0, t)\mathbf{s}||\right] \tag{4.202}$$

For the further analysis it is convenient to express the norm $||\mathbf{\Theta}(\mathbf{x}_0, t)\mathbf{s}||$ in terms of

$$\|\mathbf{\Theta}(\mathbf{x}_0, t)\mathbf{s}\| \leq \|\mathbf{\Theta}(\mathbf{X}_0, t)\| \cdot \|\mathbf{s}\| = \|\mathbf{\Theta}(\mathbf{X}_0, t)\| \quad (4.203)$$

Finally, this result is used in calculating the Lyapunov exponent according to Eq. 4.201 by using a matrix norm equal to the eigenvalue μ_{max} of $\mathbf{\Theta}(\mathbf{X}_0, t)$ with the maximum absolute value. The time domain t has to be taken large enough that the Lyapunov exponent convergences to a stationary value. For the statistical estimation of the convergence of the Lyapunov exponent, Eq. 4.202 is suitable.

4.8.3 *Linear stability analysis*

The Lyapunov exponent for the stability of the second moments of a linearized reference solution can be determined by the Itô analysis. The nonlinear stiffness matrix in Eq. 4.195 can be expanded into an asymptotic series with respect to a static loading condition. Under the assumption that the loading vector $\mathbf{F}(t)$ can be expressed in terms of a single scalar load process $f(t)$, i.e. $\mathbf{F}(\mathbf{t}) = \mathbf{F}_0 f(t)$ and that the fluctuating part is small, this series can be truncated after the linear term

$$\mathbf{M}\ddot{\mathbf{Y}} + \mathbf{C}\dot{\mathbf{Y}} + [\mathbf{K}(\mathbf{X}_{stat}) + f(t)\mathbf{K}_1]\mathbf{Y} = 0 \quad (4.204)$$

This equation of motion is projected into a subspace of dimension m and then transformed into its state space description analogous to Eq. 4.200

$$\dot{\mathbf{z}} = [\mathbf{A} + \mathbf{B}f(t)]\mathbf{z} \quad (4.205)$$

where the coefficient matrices $\mathbf{A}$ and $\mathbf{B}$ are constant. The fluctuating part of the loading function is assumed to be Gaussian white noise. Then the Eq. 4.205 can be written as a first order stochastic differential equation Lin and Cai 1995

$$\mathrm{d}\mathbf{z} = (\mathbf{A} + \pi S_{ff}\mathbf{B})\mathbf{z}\mathrm{d}t + \sqrt{2\pi S_{ff}}\,\mathbf{B}\mathbf{z}\,\mathrm{d}B(t) \quad (4.206)$$

For this system the Lyapunov exponent λ_2 for the second moments can be easily derived by applying the Itô calculus (e.g. Lin and Cai 1995). A somewhat heuristic derivation of the differential equation of the second moments is given below. We start by expanding the increment of the matrix product $\mathbf{z}\mathbf{z}^T$ into a Taylor series of order $\mathcal{O}(\mathrm{d}t)$

$$\mathrm{d}(\mathbf{z}\mathbf{z}^T) = (\mathrm{d}\mathbf{z})\mathbf{z}^T + \mathbf{z}(\mathrm{d}\mathbf{z}^T) + (\mathrm{d}\mathbf{z})(\mathrm{d}\mathbf{z}^T) \quad (4.207)$$

The second order term is necessary because $\mathrm{d}\mathbf{z}$ contains terms of the order $\sqrt{\mathrm{d}t}$. Inserting the Itô equation (4.206) we get

$$\begin{aligned}\mathrm{d}(\mathbf{z}\mathbf{z}^T) = {} & (\mathbf{A} + \pi S_{ff}\mathbf{B})\mathbf{z}\mathbf{z}^T\mathrm{d}t + \sqrt{2\pi S_{ff}}\,\mathbf{B}\mathbf{z}\mathbf{z}^T\,\mathrm{d}B(t) \\ & + \mathbf{z}\mathbf{z}^T\left(\mathbf{A}^T + \pi S_{ff}\mathbf{B}^T\right)\mathrm{d}t + \sqrt{2\pi S_{ff}}\,\,\mathbf{z}\mathbf{z}^T\mathbf{B}^T\mathrm{d}B(t) \\ & + 2\pi S_{ff}\mathbf{B}\mathbf{z}\mathbf{z}^T\mathbf{B}^T\mathrm{d}W(t)^2 + \mathcal{O}(\sqrt{\mathrm{d}t^3})\end{aligned} \quad (4.208)$$

Upon taking expectation and omitting higher-order terms, the terms containing $dW(t)$ vanish and we retain

$$\mathbf{E}[d(\mathbf{z}\mathbf{z}^T)] = (\mathbf{A} + \pi S_{ff}\mathbf{B})\mathbf{E}[\mathbf{z}\mathbf{z}^T]dt + \mathbf{E}[\mathbf{z}\mathbf{z}^T](\mathbf{A}^T + \pi S_{ff}\mathbf{B}^T)dt + 2\pi S_{ff}\mathbf{B}\mathbf{E}[\mathbf{z}\mathbf{z}^T]\mathbf{B}^T dt \tag{4.209}$$

in which the property (4.146) of the standard Wiener process $B(t)$ has been utilized. Introducing the covariance matrix

$$\mathbf{C}_{\mathbf{zz}} = \mathbf{E}[\mathbf{z}\mathbf{z}^T] \tag{4.210}$$

and the abbreviation $\mathbf{F} = \mathbf{A} + \pi S_{ff}\mathbf{B}$ we obtain the ordinary linear differential equation

$$\dot{\mathbf{C}}_{\mathbf{zz}} = \mathbf{F}\mathbf{C}_{\mathbf{zz}} + \mathbf{C}_{\mathbf{zz}}\mathbf{F}^T + 2\pi S_{ff}\mathbf{B}\mathbf{C}_{\mathbf{zz}}\mathbf{B}^T \tag{4.211}$$

Example 4.13 (Second moment stability of a SDOF-system with parametric stiffness excitation)

Consider a simple oscillator with an equation of motion in the form

$$m\ddot{x} + c\dot{x} + k[1 + f(t)]x = 0 \tag{4.212}$$

in which $f(t)$ is white noise with spectral density S_{ff}. Obviously, this system has the trivial solution $x(t) \equiv 0$. Rewriting this in state vector form we get

$$\frac{d}{dt}\begin{bmatrix} x \\ \dot{x} \end{bmatrix} = \begin{bmatrix} 0 & 1 \\ -\frac{k}{m} & -\frac{c}{m} \end{bmatrix} \begin{bmatrix} x \\ \dot{x} \end{bmatrix} + \begin{bmatrix} 0 & 0 \\ -\frac{k}{m} & 0 \end{bmatrix} \begin{bmatrix} x \\ \dot{x} \end{bmatrix} f(t) \tag{4.213}$$

The matrices $\mathbf{A}$ and $\mathbf{B}$ as utilized in the notation of Eq. 4.205 are

$$\mathbf{A} = \begin{bmatrix} 0 & 1 \\ -\frac{k}{m} & -\frac{c}{m} \end{bmatrix}; \quad \mathbf{B} = \begin{bmatrix} 0 & 0 \\ -\frac{k}{m} & 0 \end{bmatrix} \tag{4.214}$$

and from this the Itô equation

$$d\begin{bmatrix} x \\ \dot{x} \end{bmatrix} = \begin{bmatrix} 0 & 1 \\ -\frac{k}{m} & -\frac{c}{m} \end{bmatrix} \begin{bmatrix} x \\ \dot{x} \end{bmatrix} dt + \sqrt{2\pi S_{ff}} \begin{bmatrix} 0 & 0 \\ -\frac{k}{m} & 0 \end{bmatrix} dB(t) \tag{4.215}$$

The differential equation for the second moments become

$$\frac{d}{dt}\begin{bmatrix} \sigma_X^2 & \sigma_{X\dot{X}} \\ \sigma_{X\dot{X}} & \sigma_{\dot{X}}^2 \end{bmatrix} = \begin{bmatrix} 0 & 1 \\ -\frac{k}{m} & -\frac{c}{m} \end{bmatrix} \begin{bmatrix} \sigma_X^2 & \sigma_{X\dot{X}} \\ \sigma_{X\dot{X}} & \sigma_{\dot{X}}^2 \end{bmatrix} + \begin{bmatrix} \sigma_X^2 & \sigma_{X\dot{X}} \\ \sigma_{X\dot{X}} & \sigma_{\dot{X}}^2 \end{bmatrix} \begin{bmatrix} 0 & -\frac{k}{m} \\ 1 & -\frac{c}{m} \end{bmatrix} + 2\pi S_{ff} \begin{bmatrix} 0 & 0 \\ -\frac{k}{m} & 0 \end{bmatrix} \begin{bmatrix} \sigma_X^2 & \sigma_{X\dot{X}} \\ \sigma_{X\dot{X}} & \sigma_{\dot{X}}^2 \end{bmatrix} \begin{bmatrix} 0 & -\frac{k}{m} \\ 0 & 0 \end{bmatrix} \tag{4.216}$$

Assembling the elements of the covariance matrix into a vector and omitting the duplicate entry $\sigma_{X\dot{X}}$, we get

$$\frac{\mathrm{d}}{\mathrm{d}t}\begin{bmatrix}\sigma_X^2\\ \sigma_{X\dot{X}}\\ \sigma_{\dot{X}}^2\end{bmatrix} = \begin{bmatrix}0 & 2 & 0\\ -\frac{k}{m} & -\frac{c}{m} & 1\\ 2\pi S_{ff}\frac{k^2}{m^2} & -\frac{2k}{m} & -\frac{2c}{m}\end{bmatrix}\begin{bmatrix}\sigma_X^2\\ \sigma_{X\dot{X}}\\ \sigma_{\dot{X}}^2\end{bmatrix} = \mathbf{H}\begin{bmatrix}\sigma_X^2\\ \sigma_{X\dot{X}}\\ \sigma_{\dot{X}}^2\end{bmatrix} \tag{4.217}$$

It is well-known that the stability of the trivial solution of this equation is guaranteed if the real parts of all eigenvalues of $\mathbf{H}$ are negative. The eigenvalue μ of $\mathbf{H}$ with largest real value is actually real. Hence the stability limit is defined by this eigenvalue becoming zero which can easily seen from the determinant of $\mathbf{H}$

$$\det(\mathbf{H}) = \frac{4k}{m}\left(\frac{k^2}{m^2}\pi S_{ff} - \frac{c}{m}\right) \tag{4.218}$$

Setting this to zero we obtain the critical spectral density as

$$S_{ff,cr} = \frac{cm}{\pi k^2} = \frac{2\zeta}{\pi\omega_0^3} \tag{4.219}$$

which corresponds to a critical noise intensity

$$D_{cr} = \frac{2cm}{k^2} = \frac{4\zeta}{\omega_0^3} \tag{4.220}$$

Example 4.14 (Second moment stability of a SDOF-system with parametric damping excitation)

Now consider a simple oscillator with an equation of motion in the form

$$m\ddot{x} + c\left[1 + f(t)\right]\dot{x} + kx = 0 \tag{4.221}$$

in which $f(t)$ again is white noise with spectral density S_{ff}. Obviously, this system also has the trivial solution $x(t) \equiv 0$. Rewriting this in state vector form we get

$$\frac{\mathrm{d}}{\mathrm{d}t}\begin{bmatrix}x\\ \dot{x}\end{bmatrix} = \begin{bmatrix}0 & 1\\ -\frac{k}{m} & -\frac{c}{m}\end{bmatrix}\begin{bmatrix}x\\ \dot{x}\end{bmatrix} + \begin{bmatrix}0 & 0\\ 0 & -\frac{c}{m}\end{bmatrix}\begin{bmatrix}x\\ \dot{x}\end{bmatrix}f(t) \tag{4.222}$$

The matrices $\mathbf{A}$ and $\mathbf{B}$ as utilized in the notation of Eq. 4.205 are

$$\mathbf{A} = \begin{bmatrix}0 & 1\\ -\frac{k}{m} & -\frac{c}{m}\end{bmatrix};\quad \mathbf{B} = \begin{bmatrix}0 & 0\\ 0 & -\frac{c}{m}\end{bmatrix} \tag{4.223}$$

and from this the Itô equation

$$\mathrm{d}\begin{bmatrix}x\\ \dot{x}\end{bmatrix} = \begin{bmatrix}0 & 1\\ -\frac{k}{m} & -\frac{c}{m} + \pi S_{ff}\frac{c^2}{m^2}\end{bmatrix}\begin{bmatrix}x\\ \dot{x}\end{bmatrix}\mathrm{d}t + \sqrt{2\pi S_{ff}}\begin{bmatrix}0 & 0\\ 0 & -\frac{c}{m}\end{bmatrix}\mathrm{d}B(t) \tag{4.224}$$

The differential equation for the second moments become

$$\frac{\mathrm{d}}{\mathrm{d}t}\begin{bmatrix}\sigma_X^2 & \sigma_{X\dot{X}}\\ \sigma_{X\dot{X}} & \sigma_{\dot{X}}^2\end{bmatrix} = \begin{bmatrix}0 & 1\\ -\frac{k}{m} & -\frac{c}{m}+\pi S_{ff}\frac{c^2}{m^2}\end{bmatrix}\begin{bmatrix}\sigma_X^2 & \sigma_{X\dot{X}}\\ \sigma_{X\dot{X}} & \sigma_{\dot{X}}^2\end{bmatrix}$$

$$+\begin{bmatrix}\sigma_X^2 & \sigma_{X\dot{X}}\\ \sigma_{X\dot{X}} & \sigma_{\dot{X}}^2\end{bmatrix}\begin{bmatrix}0 & -\frac{k}{m}\\ 1 & -\frac{c}{m}+\pi S_{ff}\frac{c^2}{m^2}\end{bmatrix}$$

$$+\,2\pi S_{ff}\begin{bmatrix}0 & 0\\ 0 & -\frac{c}{m}\end{bmatrix}\begin{bmatrix}\sigma_X^2 & \sigma_{X\dot{X}}\\ \sigma_{X\dot{X}} & \sigma_{\dot{X}}^2\end{bmatrix}\begin{bmatrix}0 & 0\\ 0 & -\frac{c}{m}\end{bmatrix} \tag{4.225}$$

Assembling the elements of the covariance matrix into a vector we get

$$\frac{\mathrm{d}}{\mathrm{d}t}\begin{bmatrix}\sigma_X^2\\ \sigma_{X\dot{X}}\\ \sigma_{\dot{X}}^2\end{bmatrix} = \begin{bmatrix}0 & 2 & 0\\ -\frac{k}{m} & -\frac{c}{m}+2\pi S_{ff}\frac{c^2}{m^2} & 1\\ 0 & -\frac{2k}{m} & -\frac{2c}{m}+4\pi S_{ff}\frac{c^2}{m^2}\end{bmatrix}\begin{bmatrix}\sigma_X^2\\ \sigma_{X\dot{X}}\\ \sigma_{\dot{X}}^2\end{bmatrix}$$

$$= \mathbf{H}\begin{bmatrix}\sigma_X^2\\ \sigma_{X\dot{X}}\\ \sigma_{\dot{X}}^2\end{bmatrix} \tag{4.226}$$

The determinant of $\mathbf{H}$ is

$$\det(\mathbf{H}) = \frac{4k}{m}\left(\frac{2c^2}{m^2}\pi S_{ff} - \frac{c}{m}\right) \tag{4.227}$$

and from this we obtain the critical spectral density as

$$S_{ff,cr} = \frac{m}{2\pi c} = \frac{1}{4\pi\zeta\omega_0} \tag{4.228}$$

The sample stability (almost sure stability) of linear systems can be investigated by studying the logarithm of a suitably defined norm $\|z\|$. One possible definition for such a norm is

$$\|\mathbf{z}\| = \sqrt{\mathbf{z}^T\mathbf{M}\mathbf{z}} \tag{4.229}$$

in which $\mathbf{M}$ is a positive definite matrix. For the logarithm of norm, the Itô equation is (Bucher 1990)

$$\mathrm{d}\log\|\mathbf{z}\| = \frac{1}{2\|\mathbf{z}\|^2}\mathbf{z}^T(\mathbf{F}^T\mathbf{M}+\mathbf{M}\mathbf{F}+2\pi S_{ff}\mathbf{B}^T\mathbf{M}\mathbf{B})\mathbf{z}\mathrm{d}t$$

$$-\frac{1}{4\|\mathbf{z}\|^4}2\pi S_{ff}[\mathbf{z}^T(\mathbf{B}^T\mathbf{M}+\mathbf{M}\mathbf{B})\mathbf{z}]^2\mathrm{d}t$$

$$+\frac{1}{2\|\mathbf{z}\|^2}\sqrt{2\pi S_{ff}}\mathbf{z}^T(\mathbf{B}^T\mathbf{M}+\mathbf{M}\mathbf{B})\mathbf{z}\mathrm{d}B(t) \tag{4.230}$$

The Lyapunov exponent λ deciding on stability is obtained by taking expectation

$$\lambda = \mathbf{E}\left[\frac{d\log\|\mathbf{z}\|}{dt}\right] = \frac{1}{2}\mathbf{E}\left[\frac{\mathbf{z}^T(\mathbf{F}^T\mathbf{M}+\mathbf{M}\mathbf{F}+2\pi S_{ff}\mathbf{B}^T\mathbf{M}\mathbf{B})\mathbf{z}}{\mathbf{z}^T\mathbf{M}\mathbf{z}}\right] - \frac{\pi S_{ff}}{2}\mathbf{E}\left[\left(\frac{\mathbf{z}^T(\mathbf{B}^T\mathbf{M}+\mathbf{M}\mathbf{B})\mathbf{z}}{\mathbf{z}^T\mathbf{M}\mathbf{z}}\right)^2\right] \tag{4.231}$$

The first expectation can be computed easily if the matrix $\mathbf{M}$ defining the norm is chosen to be a solution to the matrix eigenvalue problem

$$\mathbf{F}^T\mathbf{M}+\mathbf{M}\mathbf{F}+2\pi S_{ff}\mathbf{B}^T\mathbf{M}\mathbf{B}-\mu\mathbf{M}=0 \tag{4.232}$$

with a real-valued eigenvalue μ. Then we obtain

$$\lambda = \frac{\mu}{2} - \frac{\pi S_{ff}}{2}\mathbf{E}\left[\left(\frac{\mathbf{z}^T(\mathbf{B}^T\mathbf{M}+\mathbf{M}\mathbf{B})\mathbf{z}}{\mathbf{z}^T\mathbf{M}\mathbf{z}}\right)^2\right] \tag{4.233}$$

Since the last term to be subtracted is positive, the quantity $\frac{\mu}{2}$ provides an upper bound for the Lyapunov exponent. In order to compute the second expectation, it is necessary to know the joint probability density function of the components of $\mathbf{z}$. In Bucher 1990 it has been suggested to follow a second-moment-based approach. The covariance matrix is governed by the differential equation (4.211) which is associated with the eigenvalue problem

$$\mathbf{F}\mathbf{C}+\mathbf{C}\mathbf{F}^T+2\pi S_{ff}\mathbf{B}\mathbf{C}\mathbf{B}^T-\mu\mathbf{C}=0 \tag{4.234}$$

in which we choose the eigenvalue μ that has the largest real part (incidentally, this eigenvalue is identical to the one given by Eq. 4.232). As time tends to infinity, the covariance matrix will be dominated by this eigenvalue and the associated eigenmatrix $\mathbf{C}$. In the next step, the Cholesky decomposition of $\mathbf{C}$

$$\mathbf{C}=\mathbf{L}\mathbf{L}^T \tag{4.235}$$

is used to construct a linear transformation into the space of uncorrelated variables $\mathbf{u}$

$$\mathbf{z}=\mathbf{L}\mathbf{u} \tag{4.236}$$

Finally, it is assumed that the probability density function of the components of $\mathbf{u}$ is rotationally symmetric such that

$$f_{\mathbf{U}}(\mathbf{u}) = g(\mathbf{u}^T\mathbf{u}) \tag{4.237}$$

It can easily be seen that the specific type of function $g(.)$ is immaterial for the second expectation in Eq. 4.233, hence for the sake of simplicity, a Gaussian distribution can be chosen. The expected value can then be computed using a simple Monte Carlo simulation scheme.

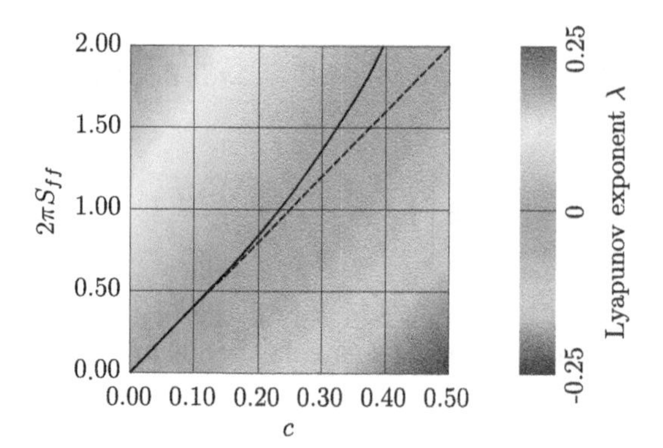

Figure 4.33 Lyapunov exponent for SDOF oscillator with random stiffness.

Example 4.15 (Sample stability of a SDOF-system with parametric stiffness excitation)
Consider again a simple oscillator with an equation of motion in the form

$$m\ddot{x} + c\dot{x} + k[1 + f(t)]x = 0 \tag{4.238}$$

in which $m = 1$, $k = 1$, and $f(t)$ is white noise with spectral density S_{ff}. The resulting Lyapunov exponent as a function of c and S_{ff} is shown in Fig. 4.33.

Stochastic averaging

The stochastic averaging method has been extensively investigated in the literature, e.g. in Lin and Cai 1995. The assumptions regarding the probability distribution are such that the joint probability density function of the displacement and the velocity is rotationally symmetric after applying a transformation of the form of Eq. 4.235. Specific to the stochastic averaging method, **L** is computed as Cholesky factor of the matrix

$$\mathbf{C} = \begin{bmatrix} \frac{k}{m} & 0 \\ 0 & 1 \end{bmatrix} = \begin{bmatrix} \omega_0^2 & 0 \\ 0 & 1 \end{bmatrix} \tag{4.239}$$

This is equivalent to the assumption

$$z_1 = x = A\cos\varphi; \quad z_2 = \dot{x} = \omega_0 A \sin\varphi \tag{4.240}$$

with φ being uniformly distributed over the unit sphere and A having an arbitrary distribution. Actually, this assumption coincides with the covariance matrix **C** obtained from the solution of Eq. 4.234 for the case of zero damping, i.e. $c=0$. The norm $\|\mathbf{z}\|$ to be used is chosen according to Eq. 4.229 with the specific matrix $\mathbf{M}$ given by

$$\mathbf{M} = \mathbf{C} = \begin{bmatrix} \omega_0^2 & 0 \\ 0 & 1 \end{bmatrix} \tag{4.241}$$

With this norm, we have

$$\|\mathbf{z}\|^2 = z_1^2\omega_0^2 + z_2^2 = A^2\omega_0^2(\cos^2\varphi + \sin^2\varphi) = A^2\omega_0^2 \tag{4.242}$$

For the case of the oscillator with random stiffness as described in Example 4.13, we have the matrix products

$$\mathbf{F}^T\mathbf{M} + \mathbf{M}\mathbf{F} + 2\pi S_{ff}\mathbf{B}^T\mathbf{M}\mathbf{B} = \begin{bmatrix} 2\pi S_{ff}\frac{k^2}{m^2} & 0 \\ 0 & -\frac{2c}{m} \end{bmatrix} \tag{4.243}$$

and

$$\mathbf{B}^T\mathbf{M} + \mathbf{M}\mathbf{B} = \begin{bmatrix} 0 & -\frac{k}{m} \\ -\frac{k}{m} & 0 \end{bmatrix} \tag{4.244}$$

The expression for the Lyapunov exponent then becomes

$$\lambda = \frac{1}{2}\mathbf{E}\left[\frac{2\pi S_{ff}\omega_0^4 A^2\cos^2\varphi - \frac{2c}{m}A^2\omega_0^2\sin^2\varphi}{A^2\omega_0^2}\right] - \frac{\pi S_{ff}}{2}\mathbf{E}\left[\left(\frac{2\omega_0^2 A\cos\varphi\,\omega_0 A\sin\varphi}{A^2\omega_0^2}\right)^2\right] \tag{4.245}$$

For a random variable φ which is uniformly distributed in the interval $[0, 2\pi)$ we can easily compute the expected values as required above

$$\mathbf{E}[\cos^2\varphi] = \int_0^{2\pi}\frac{\cos^2\varphi}{2\pi}\mathrm{d}\varphi = \frac{1}{2}; \quad \mathbf{E}[\sin^2\varphi] = \int_0^{2\pi}\frac{\sin^2\varphi}{2\pi}\mathrm{d}\varphi = \frac{1}{2}$$
$$\mathbf{E}[\cos^2\varphi\sin^2\varphi] = \int_0^{2\pi}\frac{\cos^2\varphi\sin^2\varphi}{2\pi}\mathrm{d}\varphi = \frac{1}{8} \tag{4.246}$$

Hence the Lyapunov exponent becomes

$$\lambda = \frac{1}{2}\frac{2\pi S_{ff}\omega_0^2 - \frac{2c}{m}}{2} - \frac{\pi S_{ff}}{2}\frac{4}{8} = -\frac{c}{2m} + \frac{\pi S_{ff}k}{4m} \tag{4.247}$$

From this, the critical level of noise leading to Lyapunov exponent of zero becomes

$$D_{cr} = 2\pi S_{ff,cr} = \frac{4c}{k} = \frac{8\zeta}{\omega_0} \tag{4.248}$$

The critical noise intensity D_{cr} depending on the damping ratio ζ as obtained from this equation is shown as dashed line in Fig. 4.33.

Summary

In this chapter, the properties of random processes were defined in terms of mean values, auto-covariance functions and power spectral densities. As a special case, Markov processes were discussed in some detail. The response of structural systems to random excitation was presented in terms of single- and multi-degree of freedom systems. The power spectral method was derived. As a simplification of the response analysis, the notion of white noise excitation was introduced. For multi-degree of freedom systems this allows the application of covariance analysis which was presented for both white and filtered white noise. The first passage problem was treated using approximations obtained from upcrossing rate. Monte Carlo simulation procedures was discussed together with numerical time integration methods suitable for nonlinear systems. Direct determination of the response probability density based on the solution of the Fokker-Plank-equation was presented. As an approximation method, the concept of statistical linearization was presented. The basic concepts for stochastic stability were introduced and applied to linear and nonlinear systems.

Chapter 5

Response analysis of spatially random structures

ABSTRACT: Environmental loads (e.g. wind loads) on structures exhibit random spatial and temporal fluctuations. While the temporal fluctuations can be suitably described in terms of random processes (as shown in Chapter 4), the spatial fluctuations are better described by random fields. The same type of description is useful for spatially correlated structural properties such as strength or geometry data.

This chapter deals with the description of random fields, their representation in terms of discrete random variables and methods to reduce the complexity of their description. Subsequently, the implementation of the discrete models in the context of finite element analysis is discussed and explained.

Finally, several approaches to compute statistical descriptions of the random structural response are shown and applied to several numerical examples.

5.1 Representation of random fields

5.1.1 *Basic definitions*

Let us denote by $\mathbf{x} = (x_1, x_2, \ldots, x_n)$ the structural location (e.g., distance from one support) at which we measure the value of a structural parameter (e.g., the value of the beam cross section $A(\mathbf{x})$). If we perform such a measurement for different beams (i.e., different realizations of the beam) we will observe that the value of the cross section A at location $\mathbf{x}$ will randomly vary from one measurement to the next. In other words, the cross section $A(\mathbf{x})$ at $\mathbf{x}$ is a random variable. If we additionally measure the beam cross section $A(\mathbf{y})$ at a second location $\mathbf{y} = (y_1, y_2, \ldots, y_n)$ we will observe a similar situation, i.e., the measurements will again vary randomly from one realization to the next. The cross section $A(\mathbf{y})$ at $\mathbf{y}$ is also a random variable. If we measure the values of the cross sections at both locations $\mathbf{x}$ and $\mathbf{y}$ we observe, in general, that their values will be different and that this difference will also vary randomly from one realization of the beam to the next. However, we can also observe that values from adjacent loacations do not differ as much as values that are measured at locations

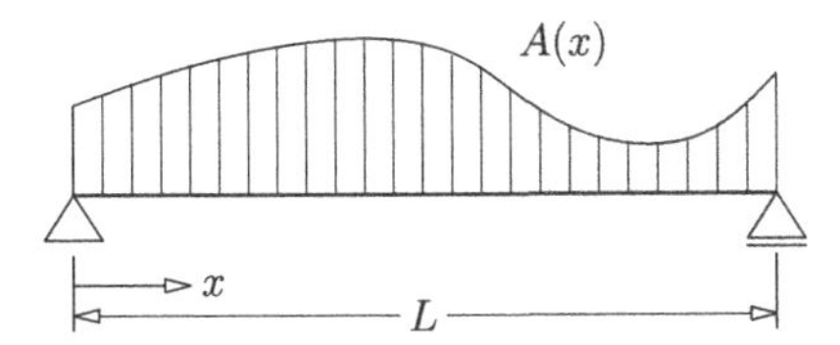

Figure 5.1 Sample function of one-dimensional random field.

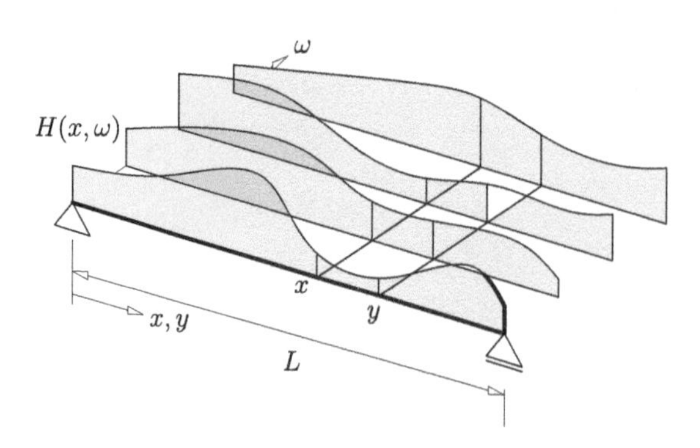

Figure 5.2 Ensemble of realizations of one-dimensional random field.

further apart. Such a behavior is an example of a covariance structure, many different types of which may be modeled by random fields.

Now let us define more precisely what we mean by random fields. A *random field* $H(\mathbf{x})$ is a real-valued random variable whose statistics (mean value, standard deviation, etc.) may be different for each value of $\mathbf{x}$ (Matthies, Brenner, Bucher, and Soares 1997; Matthies and Bucher 1999), i.e.,

$$H \in \mathbb{R}; \quad \mathbf{x} = [x_1, x_2, \ldots x_n]^T \in \mathcal{D} \subset \mathbb{R}^n \tag{5.1}$$

The *mean value function* is defined as

$$\bar{H}(\mathbf{x}) = \mathbf{E}[H(\mathbf{x})] \tag{5.2}$$

whereby the expectation operator $\mathbf{E}$ is to be taken at a fixed location $\mathbf{x}$ across the ensemble, i.e., over all possible realizations $H(\mathbf{x}, \omega)$ of the random field (see Fig. 5.2).

The spatial correlation, i.e. the fact that we observe a specific dependency structure of random field values $H(\mathbf{x})$ and $H(\mathbf{y})$ taken at different locations $\mathbf{x}$ and $\mathbf{y}$ is described by the *auto-covariance function*

$$C_{HH}(\mathbf{x}, \mathbf{y}) = \mathbf{E}[\{H(\mathbf{x}) - \bar{H}(\mathbf{x})\}\{H(\mathbf{y}) - \bar{H}(\mathbf{y})\}] \tag{5.3}$$

With respect to the form of the auto-covariance function we can classify the random fields. A random field $H(\mathbf{x})$ is called *weakly homogeneous* if

$$\bar{H}(\mathbf{x}) = \text{const.} \quad \forall \mathbf{x} \in \mathcal{D}; \quad C_{HH}(\mathbf{x}, \mathbf{x} + \boldsymbol{\xi}) = C_{HH}(\boldsymbol{\xi}) \quad \forall \mathbf{x} \in \mathcal{D} \tag{5.4}$$

This property is equivalent to the stationarity of a random process. If the covariance function depends on the distance only (not on the direction), i.e.

$$C_{HH}(\mathbf{x}, \mathbf{x} + \boldsymbol{\xi}) = C_{HH}(\|\boldsymbol{\xi}\|) \quad \forall \mathbf{x} \in \mathcal{D} \tag{5.5}$$

then a homogeneous random field $H(\mathbf{x})$ is called *isotropic*. As an example, fiber-reinforced materials with a predominant orientation of the fibers are non-isotropic in

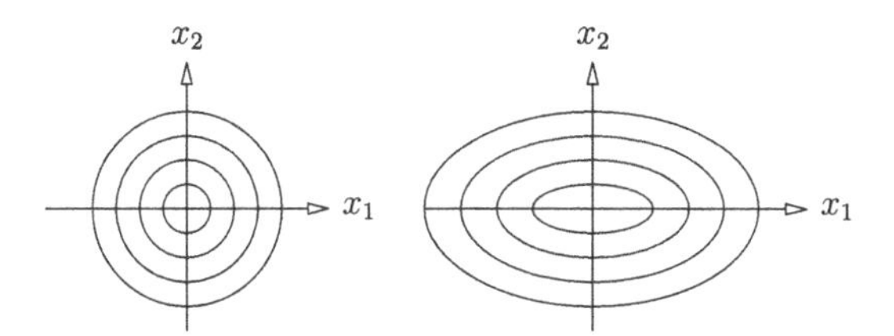

Figure 5.3 Isotropic and non-isotropic correlation contours.

their mechanical properties. This carries over to the statistical properties as well. In such a case, the contour-lines of constant correlation are elongated in one direction (cf. Fig. 5.3).

5.1.2 *Properties of the auto-covariance function*

The auto-covariance function has some mathematical properties which are important for the applicability of certain numerical methods.

Symmetry

$$C_{HH}(\mathbf{x}, \mathbf{y}) = C_{HH}(\mathbf{y}, \mathbf{x}) \tag{5.6}$$

This property is obvious from the definition of the auto-covariance function due to the commutative property of the multiplication.

Positive Definiteness

This property can be described by

$$\iint_{\mathcal{D}\otimes\mathcal{D}} w(\mathbf{x}) C_{HH}(\mathbf{x}, \mathbf{y}) w(\mathbf{y}) \, \mathrm{d}\mathbf{x} \, \mathrm{d}\mathbf{y} \geq 0 \tag{5.7}$$

which must be true for arbitrary (also non-positive) functions $w(.)$. This is equivalent to the condition

$$\sum_{k=1}^{m} \sum_{j=1}^{m} a_k C_{HH}(\mathbf{x}_k, \mathbf{x}_j) a_j \geq 0 \quad \forall a_k \in \mathbb{R}, \forall m \in \mathbb{N} \tag{5.8}$$

An interesting property of the autocovariance function of homogeneous, isotropic fields can be derived from positive definiteness (VanMarcke 1983). Let us assume that in n-dimensional space we have $m = n + 1$ points located at the vertices of an equilateral simplex (e.g. a triangle in 2D-space, cf. Fig. 5.4). All the mutual distances of these points will then be equal to, say, r. If we choose all coefficients a_k in (5.8)

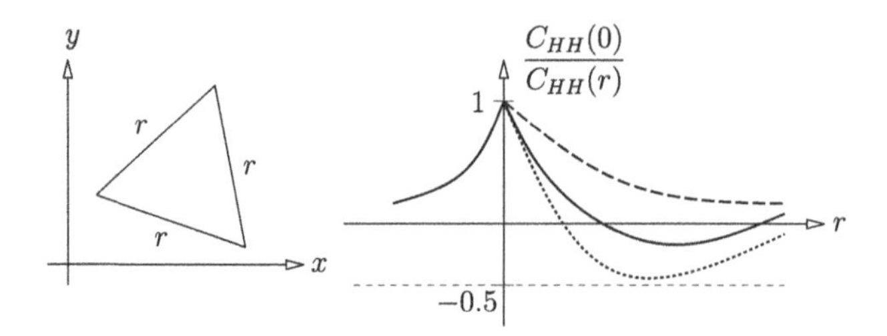

Figure 5.4 Lower Limit for Covariance Function in 2D.

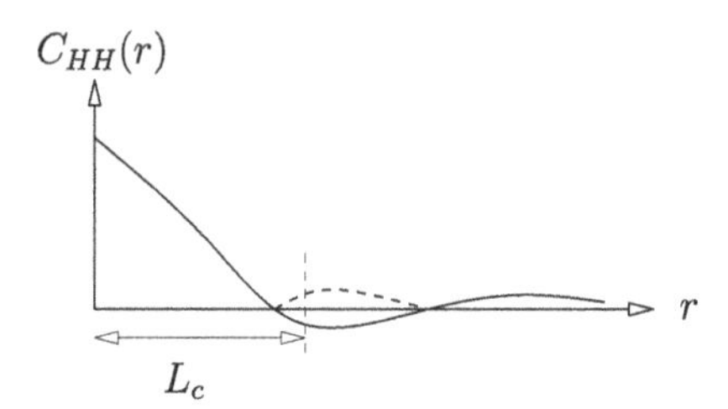

Figure 5.5 Covariance function and correlation length.

equal to 1, then we get

$$\sum_{k=1}^{m}\sum_{j=1}^{m} a_k a_j C_{HH}(\mathbf{x}_k, \mathbf{x}_j) = mC_{HH}(0) + (m^2 - m)C_{HH}(r) > 0$$

$$\rightarrow C_{HH}(r) > -\frac{C_{HH}(0)}{n} \tag{5.9}$$

This means that in 2D, the autocovariance function cannot have negative values less than $-\frac{\sigma_H^2}{2}$ and in 3D it cannot be less than $-\frac{\sigma_H^2}{3}$.

Correlation Length (isotropic case)

Define the separation distance r between two points by $r = ||\mathbf{x} - \mathbf{y}||$. Then the correlation length L_c is given by

$$L_c = \frac{\int_0^\infty r|C_{HH}(r)|\, dr}{\int_0^\infty |C_{HH}(r)|\, dr} \tag{5.10}$$

The limit $L_c \rightarrow \infty$ produces a random field which is fully correlated in the entire domain of definition thus actually describes a single random variable. The opposite limit $L_c \to 0$ produces a "finite-power white noise", i.e. a random field without any spatial correlation.

Example 5.1 (Exponential Correlation)
Let a covariance function be defined by

$$C_{HH}(r) = \sigma_H^2 \exp\left(-\frac{r}{b}\right) \tag{5.11}$$

Compute the correlation length.
The integrals required, according to (5.10), are computed as

$$\begin{aligned}\int_0^\infty r \exp\left(-\frac{r}{b}\right) \mathrm{d}r &= -br \exp\left(-\frac{r}{b}\right)\Big|_0^\infty + b \int_0^\infty \exp\left(-\frac{r}{b}\right) \mathrm{d}r = \\ &= -b^2 \exp\left(-\frac{r}{b}\right)\Big|_0^\infty = b^2\end{aligned} \tag{5.12}$$

$$\int_0^\infty \exp\left(-\frac{r}{b}\right) \mathrm{d}r = b \tag{5.13}$$

So we have $L_c = b$.

5.1.3 *Spectral decomposition*

For numerical computations it is useful to represent a continuous random field $H(\mathbf{x})$ in terms of discrete random variables c_k; $k = 1 \ldots \infty$ (Ghanem and Spanos 1991; Brenner and Bucher 1995)

$$H(\mathbf{x}) = \sum_{k=1}^{\infty} c_k \phi_k(\mathbf{x}), \quad \mathbf{x} \in \mathcal{D} \subset \mathbb{R}^n; c_k, \phi_k \in \mathbb{R} \tag{5.14}$$

The functions $\phi_k(\mathbf{x})$ are deterministic spatial shape functions which are usually chosen to represent an orthonormal basis on $\mathcal{D}$. The random coefficients c_k can be made uncorrelated, which is an extension of orthogonality into the random variable case.

This representation is usually called *Karhunen-Loéve expansion*. It is based on the following decomposition of the covariance function

$$C_{HH}(\mathbf{x}, \mathbf{y}) = \sum_{k=1}^{\infty} \lambda_k \phi_k(\mathbf{x}) \phi_k(\mathbf{y}) \tag{5.15}$$

in which λ_k and $\phi_k(\mathbf{x})$ are the eigenvalues and eigenfunctions respectively. These are solutions to the integral equation

$$\int_{\mathcal{D}} C_{HH}(\mathbf{x}, \mathbf{y}) \phi_k(\mathbf{x}) \mathrm{d}\mathbf{x} = \lambda_k \phi_k(\mathbf{y}) \tag{5.16}$$

Mathematically, Eq. (5.16) is a Fredholm integral equation of the second kind.

In most Finite-Element applications the random field $H(\mathbf{x})$ is discretized right from the start as

$$H_i = H(\mathbf{x}_i); \quad i = 1 \dots N \tag{5.17}$$

A spectral representation for the discretized random field is then obtained by

$$H_i = \sum_{k=1}^{N} \phi_k(\mathbf{x}_i) c_k = \sum_{k=1}^{N} \phi_{ik} c_k \tag{5.18}$$

Obviously, this is a matrix-vector multiplication

$$\mathbf{H} = \mathbf{\Phi c} \tag{5.19}$$

The orthogonality condition for the columns of $\mathbf{\Phi}$ becomes

$$\mathbf{\Phi}^T \mathbf{\Phi} = \mathbf{I} \tag{5.20}$$

and the covariance matrix of the components of the coefficient vector $\mathbf{c}$ is

$$\mathbf{C}_{\mathbf{cc}} = \operatorname{diag}(\sigma_{c_k}^2) \tag{5.21}$$

Both conditions can be met if the columns ϕ_k of the matrix $\mathbf{\Phi}$ solve the following eigenvalue problem

$$\mathbf{C}_{\mathbf{HH}} \phi_k = \sigma_{c_k}^2 \phi_k; \quad k = 1 \dots N \tag{5.22}$$

Statistically, the Karhunen-Loeve expansion is equivalent to a representation of the random field by means of a Principal Component Analysis (PCA).

Example 5.2 (One-dimensional random field)

Consider a one-dimensional random field $H(x)$ defined on the interval $\mathcal{I} = [0, 1]$ with an exponential correlation function $R_{HH} = \exp(-|x - y|/0.5)$, i.e. the correlation length is half of the interval length. The interval is discretized in 100 equally spaced points for which the covariance matrix is computed. The eigenvalues $\sigma_{c_k}^2$, sorted by decreasing magnitude, are shown in Fig. 5.6. It is clearly seen that the magnitude drops rapidly with increasing k. The eigenvectors ϕ_1 to ϕ_5 are also shown. Here it can be seen that the eigenvectors become more oscillatory as k increases.

5.1.4 *Conditional random fields*

There are engineering applications in which the values of a structural property are known (e.g. from measurements) in certain selected locations. In geotechnical applications this may be a specific soil property which can be determined through bore holes. Between these locations, however, a random variability is assumed.

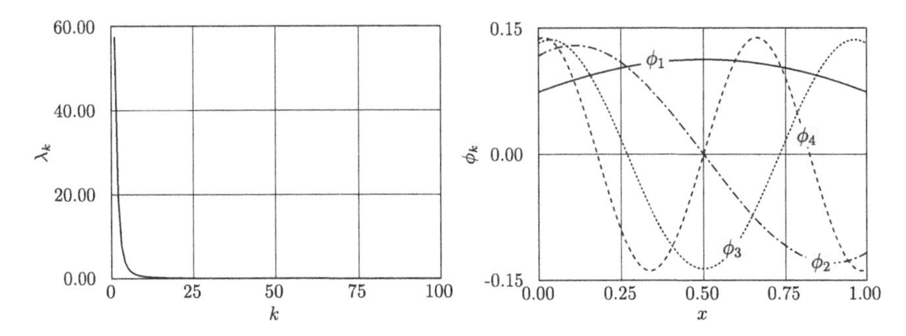

Figure 5.6 Discrete Karhunen-Loeve expansion, eigenvalues λ_k (left) and eigenvectors ϕ_k (right).

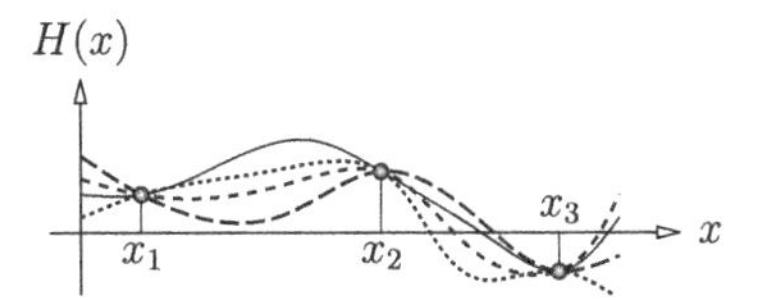

Figure 5.7 Conditional random field.

The strategy to deal with this relies on a regression approach (Ditlevsen 1991). First we assume that the structural property under consideration (without any measurements) can be modeled by a zero mean random field $H(\mathbf{x})$. This field is modified into $\hat{H}(\mathbf{x})$ by taking into account the additional knowledge.

Assume that the values of the random field $H(\mathbf{x})$ are known at the locations $\mathbf{x}_k$, $k = 1 \dots m$. We then write a stochastic interpolation for the conditional random field

$$\hat{H}(\mathbf{x}_i) = a(\mathbf{x}) + \sum_{k=1}^{m} b_k(\mathbf{x}) H(\mathbf{x}_k) \tag{5.23}$$

in which $a(\mathbf{x})$ and $b_k(\mathbf{x})$ are random interpolating functions whose statistics have yet to be determined. They are chosen to make the mean value of the difference between the random field and the conditional field zero, i.e. $\mathbf{E}[\hat{H}(\mathbf{x}) - H(\mathbf{x})] = 0$ and to minimize the variance of the difference, i.e. $\mathbf{E}[(\hat{H}(\mathbf{x}) - H(\mathbf{x}))^2] \rightarrow \text{Min}$.

Carrying out the analysis we obtain an expression for the mean value of the conditional random field.

$$\bar{\hat{H}}(\mathbf{x}) = [\mathbf{C}_{HH}(\mathbf{x}, \mathbf{x}_1)\mathbf{C}_{HH}(\mathbf{x}, \mathbf{x}_2) \dots \mathbf{C}_{HH}(\mathbf{x}, \mathbf{x}_m)]\mathbf{C}_{HH}^{-1} \begin{bmatrix} H(\mathbf{x}_1) \\ H(\mathbf{x}_2) \\ \vdots \\ H(\mathbf{x}_m) \end{bmatrix} \tag{5.24}$$

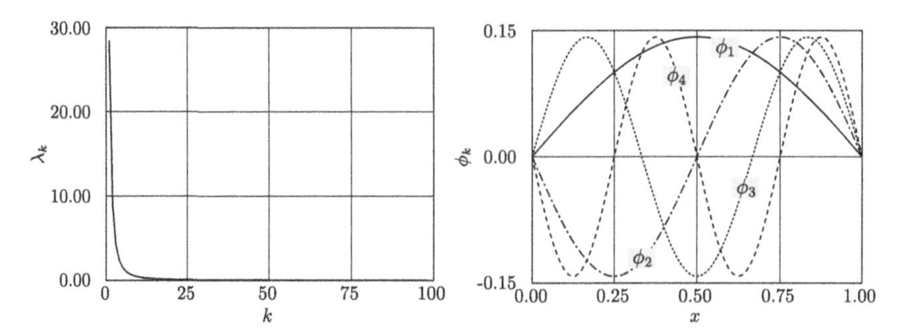

Figure 5.8 Discrete Karhunen-Loeve expansion for conditional field, eigenvalues λ_k (left) and eigenvectors ϕ_k (right).

In this equation, the matrix $\mathbf{C}_{HH}$ denotes the covariance matrix of the random field $H(\mathbf{x})$ at the locations of the measurements. The covariance matrix of the conditional random field is given by

$$\hat{\mathbf{C}}(\mathbf{x},\mathbf{y}) = \mathbf{C}(\mathbf{x},\mathbf{y}) - [\mathbf{C}_{HH}(\mathbf{x},\mathbf{x}_1)\ldots\mathbf{C}_{HH}(\mathbf{x},\mathbf{x}_m)]\mathbf{C}_{HH}^{-1}\begin{bmatrix}\mathbf{C}_{HH}(\mathbf{x}_1,\mathbf{y})\\ \vdots \\ \mathbf{C}_{HH}(\mathbf{x}_m,\mathbf{y})\end{bmatrix} \tag{5.25}$$

Example 5.3 (Conditional field on an interval)

Consider the random field like before but with the condition that the values of the field at the end of the interval should be zero. Carrying out the numerical analysis, it can be seen that now all eigenvectors, as obtained from the Karhunen-Loeve expansion, have zero values at the ends of the interval (cf. Fig. 5.8).

Example 5.4 (Conditional random field in the plane)

A zero mean random field $H(x_1,x_2)$ in the plane is described by the autocovariance function

$$C_{HH}(x_1,x_2,y_1,y_2) = \sigma_H^2 \exp\left(-\frac{|x_1-y_1]}{a} - \frac{|x_2-y_2|}{2a}\right)$$

Show that this field is homogeneous but not isotropic.

Introduce the condition that the random field should be zero at the boundaries of the square $S=[0,4a]\otimes[0,4a]$ and compute the KL-expansion for a discrete representation of the field inside the square on a grid of size a. Plot the first three eigenvectors.

The solution is carried out by arranging the discrete values of the random field in the plane into a vector. The position index is computed from the scheme as outlined in Fig. 5.9. The `octave` script is given below

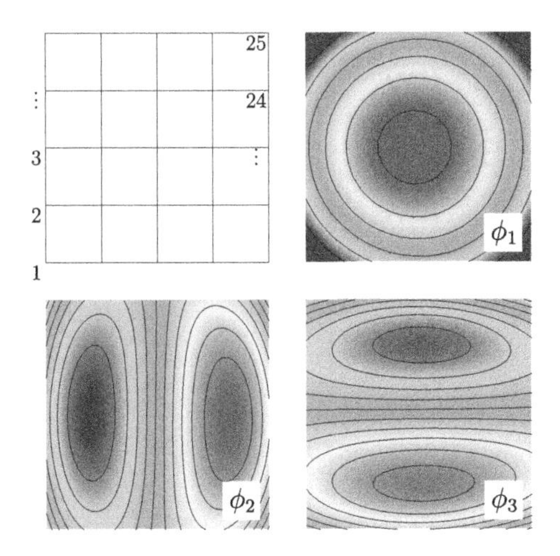

Figure 5.9 Discrete Karhunen-Loeve expansion for conditional field.

```
 1  function [v,d] = eigsort (x)
 2    length  = issquare (x);
 3    [v,d]   = eig (x);
 4    [dd,ix] = sort (-diag (d));
 5    for i=1:length
 6      d(i,i)   = - dd(i);
 7   end;
 8   v         = v(:,ix);
 9  endfunction
10  %
11  a=1;
12  ss=0.01;
13  border=[1,2,3,4,5,6,10,11,15,16,20,21,22,23,24,25];
14  inside=[7,8,9,12,13,14,17,18,19,22,23,24];
15  for i=1:25
16    x1(i)=a*floor((i-1)/5);
17    x2(i)=a*mod(i-1,5);
18  endfor
19  c=zeros(25);
20  for i=1:25
21    for k=1:25
22      c(i,k)=ss*exp(-abs(x1(i)-x1(k))/a
23        -abs(x2(i)-x2(k))/2/a);
24    endfor
25  endfor
26  chh=zeros(16);
27  for i=1:16
28    for k=1:16
```

```
29        chh(i,k)=c(border(i), border(k));
30      endfor
31    endfor
32    chhi=inv(chh);
33    cc=zeros(9);
34    for i=1:9
35      for k=1:9
36        ci=[];
37        ck=[];
38        for l=1:16
39          ci=[ci,c(inside(i),border(l))];
40          ck=[ck,c(inside(k),border(l))];
41        endfor
42     cc(i,k)=c(inside(i),inside(k))-ci*chhi*ck';
43      endfor
44    endfor
45
46    [v,l]=eigsort(cc);
47    [fid,msg]=fopen("karhunen_plate.dat", "w");
48    z=zeros(25);
49    for i=1:9
50      z(inside(i))=v(i,5);
51    endfor
52    for i=1:25
53      fprintf(fid," %g %g %g\n", x1(i), x2(i), z(i));
54    endfor
55    fclose(fid);
```

5.1.5 *Local averages of random fields*

In reality, it is usually very difficult (or, actually, conceptionally impossible) to observe structural properties in one point. Any measuring device typically gives information about averages over a certain finite domain Ω.

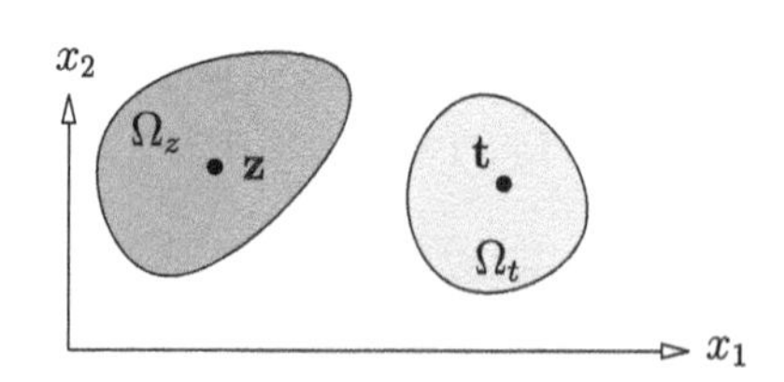

Figure 5.10 Averaging Domains for a Random Field.

Given a homogeneous random field $H(\mathbf{x})$, we define a local average of this field by

$$G_\Omega(\mathbf{z}) = \frac{1}{A_{\Omega_z}} \int\limits_{\Omega_z} H(\mathbf{x}) \mathrm{d}A \tag{5.26}$$

Here, $\mathbf{z}$ is a suitable chosen center point of Ω_z, and A_{Ω_z} is the area content of Ω_z. It is obvious that G will be smoother (less random) than H. In fact, as Ω becomes large and tends to $\mathbb{R}^2$, G will become deterministic, i.e. equal to the expected value $\bar{H}$.

The mean value of the locally averaged field can be computed from

$$\mathbf{E}[G_\Omega(\mathbf{z})] = \bar{G}_\Omega(\mathbf{z}) = \frac{1}{A_{\Omega_z}} \int\limits_{\Omega_z} \mathbf{E}[H(\mathbf{x})] \mathrm{d}A = \bar{H} \tag{5.27}$$

Its covariance function is given by

$$C_{GG}(\mathbf{z}, \mathbf{t}) = \frac{1}{A_{\Omega_z} A_{\Omega_t}} \int\limits_{\Omega_z} \int\limits_{\Omega_t} C_{HH}(\mathbf{x}, \mathbf{y}) \mathbf{dxdy} \tag{5.28}$$

Example 5.5 (Local Average of a White Noise Field)

Consider a one-dimensional zero-mean random field $H(x)$ with a covariance function

$$C_{HH}(x - y) = D_0 \delta(x - y) \tag{5.29}$$

Define the averaging domain to be an interval of length $2R$ symmetric about z

$$\Omega_z = [-R + z, R + z] \tag{5.30}$$

We then compute the covariance function for the local average

$$C_{GG}(z, t) = \frac{D_0}{4R^2} \int\limits_{-R+z}^{R+z} \int\limits_{-R+t}^{R+t} \delta(x - y) \, \mathrm{d}x\mathrm{d}y \tag{5.31}$$

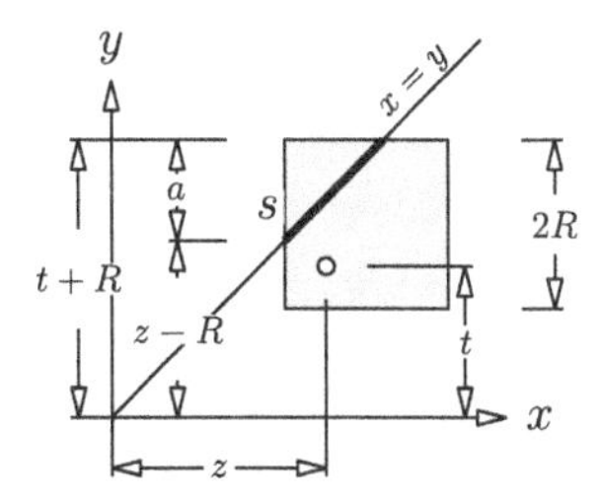

Figure 5.11 Integration for local average of white noise random field.

The location of the singularity in the $x-y$-plane is shown in the figure. The integral then evaluates to the length s of the intersection of the first median with the square. This depends on z and t. The length of a is easily found as $a=t-z+2R$ if $z>t$ and $t-z<2R$; and zero otherwise. From that we obtain

$$C_{GG}(z,t)=\begin{cases}\frac{\sqrt{2}D_0}{2R}\left(1-\frac{|z-t|}{2R}\right) & |z-t|<2R\\ 0 & \text{else}\end{cases} \tag{5.32}$$

Taking $z=t$ in this equation we obtain the variance of local average as

$$\sigma_G^2=\frac{\sqrt{2}D_0}{2R} \tag{5.33}$$

which tends to zero as the averaging length R grows.

5.2 Geometrical imperfections

In structural applications the influence of manufacturing tolerances leads to deviations of the actual geometrical shapes from those designed. Since the structural analysis is typically based on the target geometry as designed it may be unsafe to neglect the random deviations in geometry (Bucher 2006). The differences between the target geometry and the actual random geometry can be conveniently expressed as a random field involving the structural coordinates (cf. Fig. 5.12). A certain regularity or "waviness" of the geometrical imperfections can be modeled by choosing a suitable spatial correlation function. The possible effect of geometrical imperfections on the behavior of structures which are very slender and prone to stability failure will be illustrated by the following example.

Example 5.6 (Loss of stability due to geometrical imperfections)
Consider a shallow shell-like spatial structure as sketched in Fig. 5.13. This structure consists of slender beams connected rigidly, forming a doubly curved dome. The beams have a circular cross section with a diameter of 56 mm. The material is structural steel with a modulus of elasticity $E=210$ GPa and a mass density of $\rho=7800$ kg/m^3. The structure is simply supported at all nodes on the circumference. Assuming perfect geometry, the critical load multiplier λ_c for the dead load of the structure can be computed. Its numerical value is $\lambda_c=2.17$. This indicates that the geometrically perfect structure can carry more than twice its own weight. The random geometrical imperfections are represented in terms of random deviations $\hat{z}$ of the vertical coordinates z

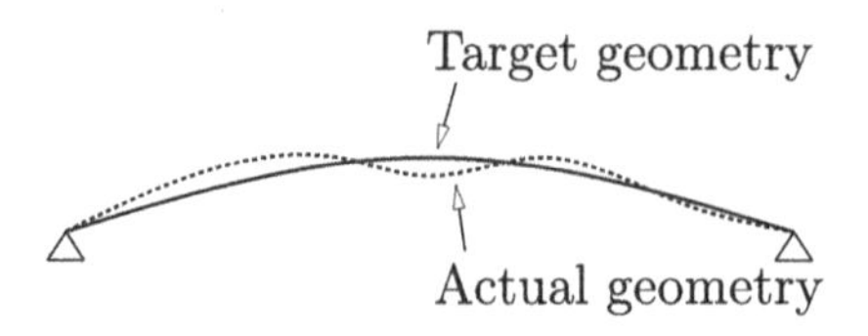

Figure 5.12 Geometrical imperfections.

of the nodes at which the beams are connected. The spatial autocovariance function for these random deviations is assumed to be of the exponential type

$$C_{\hat{z}\hat{z}} = \sigma_{\hat{z}}^2 \exp\left(-\frac{r}{L_c}\right) \tag{5.34}$$

in which r denotes the spatial sparation distance between two considered nodes, $L_c = 20\,\mathrm{m}$ is the correlation length and the standard deviation $\sigma_{\hat{z}}$ of the imperfections is varied in the analysis. The Karhunen-Loeve spectral decomposition of the random field is carried out utilizing the eigensolution of the discrete covariance matrix of the field evaluated at the nodes. The eigenvectors that belong to the four largest eigenvalues of the covariance matrix are shown in Fig. 5.14. Based on these eigenvectors, a Monte-Carlo simulation is carried out. This simulation is repeated for different magnitudes of the geometrical imperfections as expressed by $\sigma_{\hat{z}}$. In this simulation, the

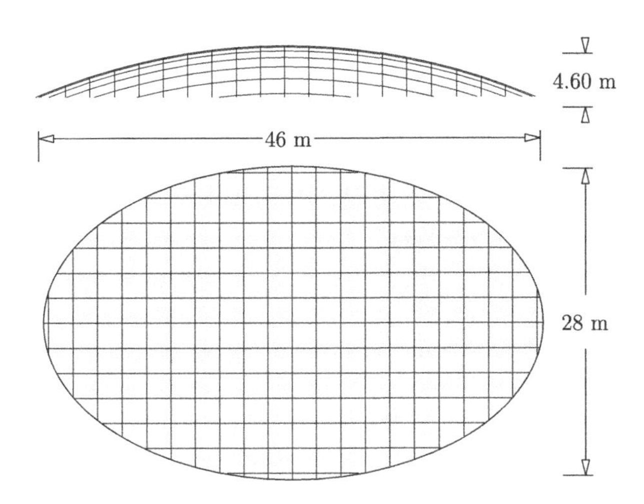

Figure 5.13 Sketch of geometrically perfect structure.

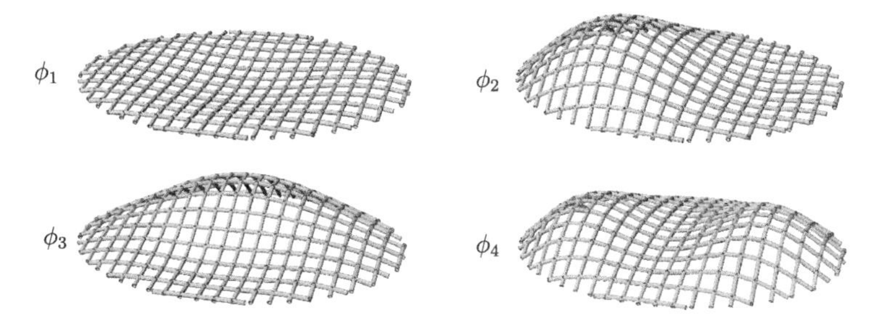

Figure 5.14 Eigenvectors of covariance matrix.

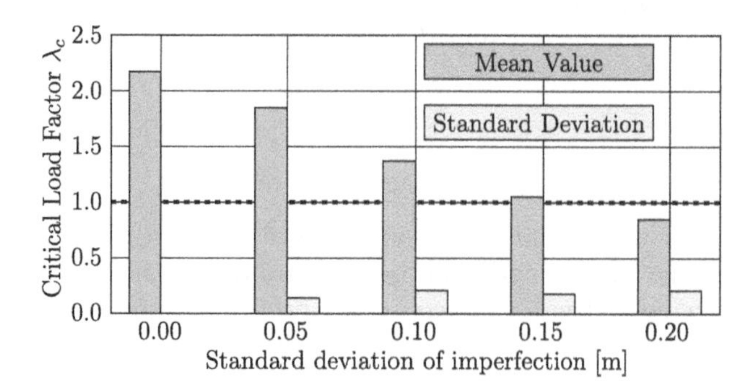

Figure 5.15 Influence of geometrical imperfections.

critical load factor is computed through linearized stability analysis using a geometrically nonlinear finite element model of the structure. The results, given in Fig. 5.15, show that imperfections of a few centimeters can reduce the load carrying capacity so significantly that even collapse of the structure under its own weight becomes highly probable.

5.3 Stochastic finite element formulation

5.3.1 *Elasticity (Plane stress)*

We consider a two-dimensional domain Ω as shown in Fig. 5.16. The boundary $\partial\Omega$ is the union of two disjoint sets G_1 and G_2. On G_1 the forces (tractions) **f** are given, on G_2 the displacements **u** are given. For simplicity, we assume that there are no body forces in the interior of Ω.

The equilibrium conditions in Ω are

$$\begin{aligned} \frac{\partial \sigma_{xx}}{\partial x} + \frac{\partial \tau_{xy}}{\partial y} &= 0 \\ \frac{\partial \tau_{xy}}{\partial x} + \frac{\partial \sigma_{yy}}{\partial y} &= 0 \end{aligned} \tag{5.35}$$

Assuming linear elasticity with elastic modulus E and Poisson's ratio ν, the stresses can be related to the strains by

$$\begin{aligned} \sigma_{xx} &= \frac{E}{(1+\nu)(1-2\nu)}[(1-\nu)\varepsilon_{xx} + \nu\varepsilon_{yy}] \\ \sigma_{yy} &= \frac{E}{(1+\nu)(1-2\nu)}[\nu\varepsilon_{xx} + (1-\nu)\varepsilon_{yy}] \\ \tau_{xy} &= \frac{E}{2(1+\nu)}\,\gamma_{xy} \end{aligned} \tag{5.36}$$

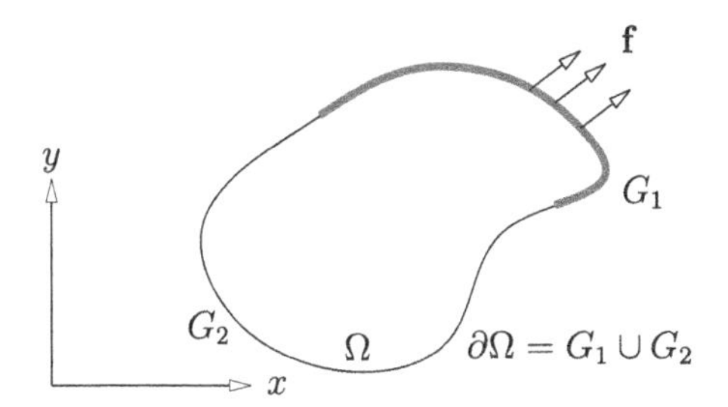

Figure 5.16 Elastic Domain with Load and Displacement Boundaries.

For convenience this is frequently written in vector-matrix notation

$$\mathbf{s} = \begin{bmatrix} \sigma_{xx} \\ \sigma_{xx} \\ \tau_{xy} \end{bmatrix} = \frac{E}{1+\nu} \begin{bmatrix} \frac{1-\nu}{1-2\nu} & \frac{\nu}{1-2\nu} & 0 \\ \frac{\nu}{1-2\nu} & \frac{1-\nu}{1-2\nu} & 0 \\ 0 & 0 & \frac{1}{2} \end{bmatrix} \begin{bmatrix} \varepsilon_{xx} \\ \varepsilon_{yy} \\ \gamma_{xy} \end{bmatrix} = \mathbf{De} \tag{5.37}$$

The strains are related to the displacements by

$$\begin{aligned} \varepsilon_{xx} &= \frac{\partial u}{\partial x} \\ \varepsilon_{yy} &= \frac{\partial v}{\partial y} \\ \gamma_{xy} &= \frac{\partial u}{\partial y} + \frac{\partial v}{\partial x} \end{aligned} \tag{5.38}$$

Note that this definition is useful for small strains only. For moderate to large strains, the definition must be augmented to include at least quadratic terms in $\frac{\partial u}{\partial x}$ etc.

The differential operators in (5.35) and (5.38) can be written as matrices

$$\mathbf{L}_1 = \begin{bmatrix} \frac{\partial}{\partial x} & 0 & \frac{\partial}{\partial y} \\ 0 & \frac{\partial}{\partial y} & \frac{\partial}{\partial x} \end{bmatrix}; \quad \mathbf{L}_2 = \begin{bmatrix} \frac{\partial}{\partial x} & 0 \\ 0 & \frac{\partial}{\partial y} \\ \frac{\partial}{\partial y} & \frac{\partial}{\partial x} \end{bmatrix} \tag{5.39}$$

so that

$$\mathbf{L}_1\mathbf{s} = 0; \quad \mathbf{L}_2 \begin{bmatrix} u \\ v \end{bmatrix} = \mathbf{L}_2\mathbf{u} = \mathbf{e} \tag{5.40}$$

Putting this together with (5.37) we obtain

$$\mathbf{L}_1\mathbf{D}\mathbf{L}_2\mathbf{u} = 0 \tag{5.41}$$

as the equilibrium condition for the interior of Ω.

For structural analysis, we are frequently interested in the relationship between the applied loads (and/or the reaction forces) f and the internal stresses s.

5.3.2 *Principle of virtual work*

In an equilibrium configuration, the virtual work done by the external forces f is equal to the virtual work done by the internal stresses s. Hence

$$\delta W = \int_V \mathbf{s}^T \delta\mathbf{e}\, \mathrm{d}V - \int_{G_1} \mathbf{f}^T \delta\mathbf{u}_1\, \mathrm{d}A = 0 \tag{5.42}$$

Here, δW is the total virtual work, $\delta\mathbf{u}_1$ is the virtual displacement field on G_1 and $\delta\mathbf{e}$ is a compatible virtual strain field in Ω.

Example 5.7 (Cantilever Rod)

Given F_1, E, A and L, compute u_1 using the principle of virtual work.

We apply a virtual displacement δu_1 at the right end of the rod and compute a compatible (linearly distributed) virtual displacement field $\delta u(x)$ in the rod. From that we obtain virtual strains by means of

$$\delta\varepsilon_{xx} = \frac{\mathrm{d}\delta u(x)}{\mathrm{d}x} = \frac{1}{L}\delta u_1 \tag{5.43}$$

The internal stress σ_{xx} is related to the strain ε_{xx} by

$$\sigma_{xx} = E\varepsilon_{xx} \tag{5.44}$$

So we obtain

$$\delta W = \int_0^L AE\varepsilon_{xx}\frac{1}{L}\delta u_1 \mathrm{d}x - F_1\delta u_1 = 0 \tag{5.45}$$

and since this equation must be true for arbitrary δu_1

$$\begin{aligned} F_1 &= \int_0^L AE\varepsilon_{xx}\frac{1}{L}\mathrm{d}x = \frac{EA}{L}\int_0^L \frac{\mathrm{d}u}{\mathrm{d}x}\mathrm{d}x = \frac{EA}{L}\left[u(L) - u(0)\right] \\ &= \frac{EA}{L}u_1 \rightarrow u_1 = \frac{F_1 L}{EA} \end{aligned} \tag{5.46}$$

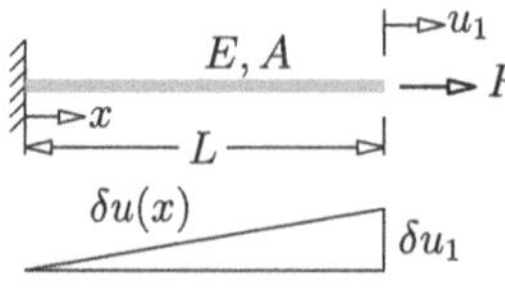

Figure 5.17 Cantilever rod.

5.3.3 *Finite element formulation*

The displacement field $\mathbf{u}(\mathbf{x})$ in a domain Ω_e is represented in terms of a finite number of nodal displacements assembled into a vector $\mathbf{U}^e$ of size n_e. The field is described by a set of interpolating functions (shape functions) $H_{ij}(\mathbf{x})$.

The interpolation is done by

$$u_i(x,y) = \sum_{j=1}^{n_e} H_{ij}(x,y) U_j^e \tag{5.47}$$

Typically, polynomials are chosen as the interpolation functions. They have to be normalized to match the conditions

$$H_{ij}(x_k, y_k) = \delta_{jk}; \quad i = 1, 2 \tag{5.48}$$

Here, δ_{jk} denotes the Kronecker Delta symbol.

The element strains are computed by taking the appropriate derivatives of (5.47)

$$\begin{aligned}
\varepsilon_{xx} &= \frac{\partial u}{\partial x} = \frac{\partial}{\partial x}\left[\sum_{j=1}^{n_e} H_{1j}(x,y) U_j^e\right] = \sum_{j=1}^{n_e} \frac{\partial H_{1j}}{\partial x} U_j^e \\
\varepsilon_{yy} &= \frac{\partial v}{\partial y} = \frac{\partial}{\partial y}\left[\sum_{j=1}^{n_e} H_{2j}(x,y) U_j^e\right] = \sum_{j=1}^{n_e} \frac{\partial H_{2j}}{\partial y} U_j^e \\
\gamma_{xy} &= \frac{\partial u}{\partial y} + \frac{\partial v}{\partial x} = \sum_{j=1}^{n_e} \left(\frac{\partial H_{1j}}{\partial y} + \frac{\partial H_{2j}}{\partial x}\right) U_j^e
\end{aligned} \tag{5.49}$$

which can be compactly written as (assuming small strains)

$$\mathbf{e} = \mathbf{B}(x,y)\mathbf{U}^e \tag{5.50}$$

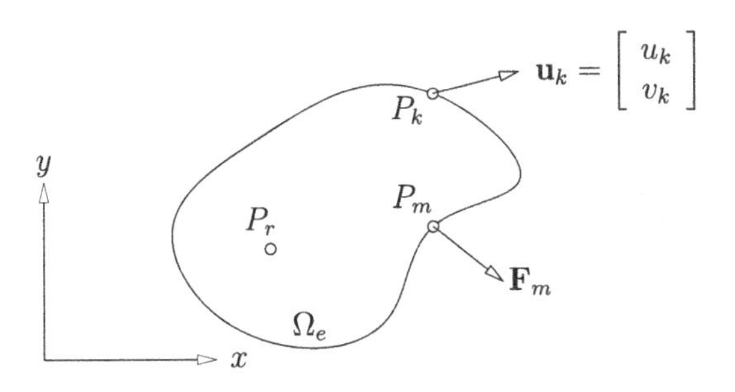

Figure 5.18 Discrete Forces and Displacements.

From this, virtual strains are easily generated

$$\delta \mathbf{e} = \frac{\partial \mathbf{e}}{\partial \mathbf{U}^e} \delta \mathbf{U}^e = \mathbf{B}(x, y) \delta \mathbf{U}^e \tag{5.51}$$

Using the principle of virtual work we can derive a relation between the element nodal displacements $\mathbf{U}^e$ and the element nodal forces $\mathbf{F}^e$

$$\delta W = \int_{\Omega_e} \mathbf{s}^T \delta \mathbf{e} \, \mathrm{d}V - \mathbf{F}^{eT} \delta \mathbf{U}^e = 0 \tag{5.52}$$

Together with (5.50) this becomes

$$\int_{\Omega_e} \mathbf{s}^T \mathbf{B} \delta \mathbf{U}^e \, \mathrm{d}V - \mathbf{F}^{eT} \delta \mathbf{U}^e = 0 \tag{5.53}$$

so that

$$\begin{aligned} \mathbf{F}^e &= \int_{\Omega_e} \mathbf{B}^T \mathbf{s} \, \mathrm{d}V = \int_{\Omega_e} \mathbf{B}^T \mathbf{D} \mathbf{e} \, \mathrm{d}V = \int_{\Omega_e} \mathbf{B}^T \mathbf{D} \mathbf{B} \mathbf{U}^e \, \mathrm{d}V \\ &= \int_{\Omega_e} \mathbf{B}^T \mathbf{D} \mathbf{B} \, \mathrm{d}V \, \mathbf{U}^e = \mathbf{K}^e \mathbf{U}^e \end{aligned} \tag{5.54}$$

The matrix $\mathbf{K}^e$ is called element stiffness matrix. The integrals appearing in this matrix are usually computed numerically, e.g. by Gauss integration.

Example 5.8 (Shape Functions and Stiffness Matrix for a Rod Element)
Consider the case of a three-node rod element as shown in Fig. 5.19.
In order to be able to satisfy the normalization conditions at three nodes, the shape functions have to be quadratic in x

$$H_{1j} = A_j + B_j x + C_j x^2 \tag{5.55}$$

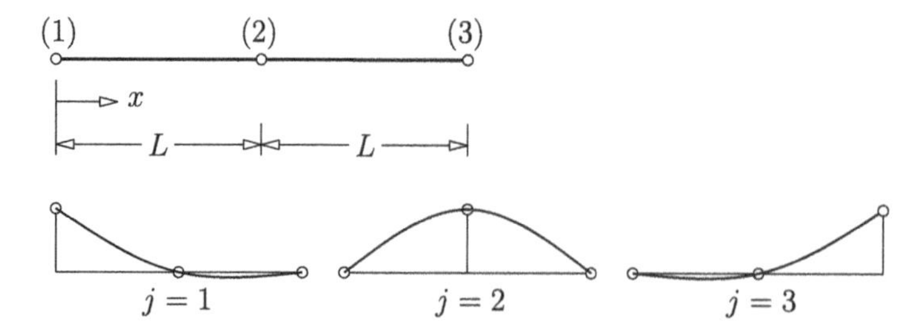

Figure 5.19 Three-node rod element and shape functions.

The coordinates of the nodes are $x_1 = 0$; $x_2 = L$; $x_3 = 2L$. The normalization conditions for the first shape function are

$$\begin{aligned} H_{11}(x_1) &= 1 = A_1 \\ H_{12}(x_2) &= 0 = A_1 + B_1 L + C_1 L^2 \\ H_{13}(x_3) &= 0 = A_1 + 2B_1 L + 4C_1 L^2 \end{aligned} \tag{5.56}$$

The solution to this system of equations is

$$A_1 = 1; \; B_1 = -\frac{3}{2L}; \; C_1 = \frac{1}{2L^2} \tag{5.57}$$

so that

$$H_{11} = 1 - \frac{3x}{2L} + \frac{x^2}{2L^2} \tag{5.58}$$

The remaining shape functions are found analogously

$$H_{12} = \frac{2x}{L} - \frac{x^2}{L^2}; \quad H_{13} = -\frac{x}{2L} + \frac{x^2}{2L^2} \tag{5.59}$$

A plot of these functions is shown in Fig. 5.19.

Since there is only uniaxial strain along the rod axis, the **B**-matrix can be reduced to a simple vector

$$\mathbf{B}(x) = \left[-\tfrac{3}{2L} + \tfrac{x}{L^2} \;\; \tfrac{2}{L} - \tfrac{2x}{L^2} \;\; -\tfrac{1}{2L} + \tfrac{x}{L^2} \right] \tag{5.60}$$

and the elasticity matrix reduces to a scalar containing the modulus of elasticity E so that

$$\mathbf{B}^T\mathbf{D}\mathbf{B} = E \cdot \begin{bmatrix} \left(\frac{x}{L^2} - \frac{3}{2L}\right)^2 & \left(\frac{2}{L} - \frac{2x}{L^2}\right)\left(\frac{x}{L^2} - \frac{3}{2L}\right) & \left(\frac{x}{L^2} - \frac{3}{2L}\right)\left(\frac{x}{L^2} - \frac{1}{2L}\right) \\ \left(\frac{2}{L} - \frac{2x}{L^2}\right)\left(\frac{x}{L^2} - \frac{3}{2L}\right) & \left(\frac{2}{L} - \frac{2x}{L^2}\right)^2 & \left(\frac{2}{L} - \frac{2x}{L^2}\right)\left(\frac{x}{L^2} - \frac{1}{2L}\right) \\ \left(\frac{x}{L^2} - \frac{3}{2L}\right)\left(\frac{x}{L^2} - \frac{1}{2L}\right) & \left(\frac{2}{L} - \frac{2x}{L^2}\right)\left(\frac{x}{L^2} - \frac{1}{2L}\right) & \left(\frac{x}{L^2} - \frac{1}{2L}\right)^2 \end{bmatrix} \tag{5.61}$$

Integration over the element volume results in

$$\mathbf{K}^e = A \int_0^{2L} \mathbf{B}^T\mathbf{D}\mathbf{B} \, dx = \frac{EA}{6L} \begin{bmatrix} 7 & -8 & 1 \\ -8 & 16 & -8 \\ 1 & -8 & 7 \end{bmatrix} \tag{5.62}$$

Exercise 5.1 (Stiffness matrix of triangular (CST) element)

Consider an equilateral triangle made of an elastic material (modulus of elasticity E, Poisson's ratio $\nu = 0$) as shown in Fig. 5.20.

The thickness of the element is t. Compute the element stiffness matrix for deformations in the $x - y$-plane.

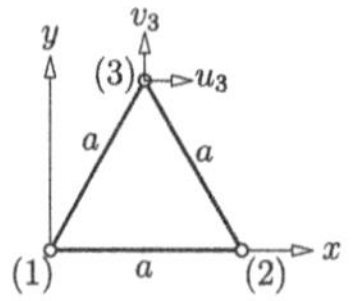

Figure 5.20 Constant strain triangle finite element.

Solution: The element stiffness matrix is given by

$$\mathbf{K}^e = \frac{Et}{a^2}\begin{bmatrix} \frac{7}{6} & -\frac{1}{2\sqrt{3}} & -\frac{7}{6} & \frac{1}{2\sqrt{3}} & \frac{1}{3} & 0 \\ -\frac{1}{2\sqrt{3}} & \frac{5}{6} & \frac{1}{2\sqrt{3}} & -\frac{5}{6} & -\frac{1}{\sqrt{3}} & \frac{2}{3} \\ -\frac{7}{6} & \frac{1}{2\sqrt{3}} & \frac{7}{6} & -\frac{1}{2\sqrt{3}} & -\frac{1}{3} & 0 \\ \frac{1}{2\sqrt{3}} & -\frac{5}{6} & -\frac{1}{2\sqrt{3}} & \frac{5}{6} & \frac{1}{\sqrt{3}} & -\frac{2}{3} \\ \frac{1}{3} & -\frac{1}{\sqrt{3}} & -\frac{1}{3} & \frac{1}{\sqrt{3}} & \frac{2}{3} & 0 \\ 0 & \frac{2}{3} & 0 & -\frac{2}{3} & 0 & \frac{4}{3} \end{bmatrix}$$

5.3.4 *Structural response*

If a structure is discretized into several finite elements, the element forces caused by the displacements are then assembled into a global restoring force vector **F** and the element nodal displacements are assembled into a global displacement vector **U**. In a static analysis, the equilibrium conditions at the nodes require that the applied load vector **P** should be equal to the restoring force vector **F**. Since all element forces $\mathbf{F}^e$ are linearly related to the respective displacements $\mathbf{U}^e$, the global vectors **F** and **U** have a linear relation as well. So the equilibrium conditions are

$$\mathbf{KU} = \mathbf{P} \tag{5.63}$$

This is a system of linear equations for the elements of the global displacement vector **U**.

Especially for the case of multiple load cases (multiple right hand sides) it is advantageous to factor the global stiffness matrix **K** into

$$\mathbf{K} = \mathbf{L\Lambda L}^T \tag{5.64}$$

in which **L** is a lower triangular matrix and **Λ** is a diagonal matrix. In this way, the explicit computation of the inverse stiffness matrix $\mathbf{K}^{-1}$ can be avoided.

In case of structural dynamics, the element assembly leads to a system of second order differential equations

$$\mathbf{M}\frac{\mathrm{d}^2}{\mathrm{d}t^2}\mathbf{U} + \mathbf{KU} = \mathbf{P}(t) \tag{5.65}$$

Here, **M** is the system mass matrix which can be obtained in a way analogous to the stiffness matrix. When applying modal analysis, first of all an eigenvalue problem needs to be solved

$$(\mathbf{K} - \omega^2\mathbf{M})\mathbf{\Phi} = 0 \tag{5.66}$$

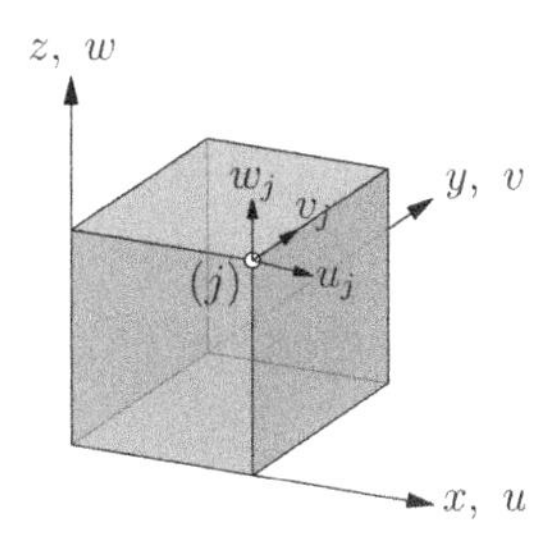

Figure 5.21 Volume element with nodal displacements.

in which the eigenvalues ω_k are the squares of the natural frequencies of the system and the eigenvectors $\mathbf{\Phi}_k$ are the modes of vibration. These modes satisfy the orthogonality relations

$$\mathbf{\Phi}_i \mathbf{M} \mathbf{\Phi}_k = \delta_{ik}; \quad \mathbf{\Phi}_i \mathbf{K} \mathbf{\Phi}_k = \omega_k^2 \delta_{ik} \tag{5.67}$$

5.3.5 *Stochastic stiffness matrix*

As discussed earlier, the element stiffness matrix $\mathbf{K}^e$ relates the nodal forces $\mathbf{F}^e$ to the nodal displacements $\mathbf{U}^e$

$$\mathbf{F}^e = \mathbf{K}^e \mathbf{U}^e \tag{5.68}$$

in which, for the element with m^e nodes $(j); j = 1 \ldots m^e$ as sketched in Fig. 5.21

$$\mathbf{F}^e = [(f_{xj}, f_{yj}, f_{zj}); j = 1 \ldots m^e]^T; \quad \mathbf{U}^e = [(u_j, v_j, w_j); j = 1 \ldots m^e]^T \tag{5.69}$$

Based on the principle of virtual work, the element stiffness matrix for a linear material law (assuming geometrical linearity as well) is obtained as

$$\mathbf{K}^e = \int_{V^e} \mathbf{B}^T(x,y) \mathbf{D}(x,y) \mathbf{B}(x,y) dV^e \tag{5.70}$$

Typically, the strain interpolation matrix $\mathbf{B}(x,y)$ is chosen in polynomial form, i.e.

$$\mathbf{B}(x,y) = \sum_{k+l \leq r} \sum \mathbf{B}_{kl} x^k y^l; \quad k, l, r \geq 0 \tag{5.71}$$

In this equation, $\mathbf{B}_{kl}$ are constant matrices. In fact, for the CST element shown in Fig. 5.20 there is only one such matrix, i.e. $\mathbf{B}_{00}$. Assuming that the system randomness is described by a random elastic modulus $E(x, y)$, the elasticity matrix $\mathbf{D}(x, y)$ can be written as

$$\mathbf{D}(x,y) = \mathbf{D}_0 E(x,y) \tag{5.72}$$

Using the polynomial form of $\mathbf{B}(x, y)$, the element stiffness matrix finally becomes

$$\mathbf{K}^e = \sum_{k+l \leq r} \sum \sum_{m+n \leq r} \sum \mathbf{B}_{kl}^T \mathbf{D}_0 \mathbf{B}_{mn} \int_{V^e} E(x,y) x^k y^l x^m y^n \, \mathrm{d}V^e \tag{5.73}$$

The last term in this equation is a so-called *weighted integral* of the random field $E(x, y)$.

$$X^e_{klmn} = \int_{V^e} E(x,y) x^k y^l x^m y^n \, \mathrm{d}V^e \tag{5.74}$$

Using this representation, it is possible to achieve a description of the random variation of the element stiffness matrix in terms of the mean values and the covariance matrix of the weighted integrals.

Example 5.9 (Three-Node Rod Element)
Consider again a three node rod element as discussed previously. The elastic modulus is assumed to be a homogeneous randon field with mean value $\bar{E}$ and an autocovariance function

$$C_{EE}(x,y) = \sigma_E^2 \frac{1}{1 + \frac{(x-y)^2}{2L^2}} \tag{5.75}$$

We will compute the weighted integral representation of the element stiffness matrix.
The **B**-matrix as derived in (5.60) is split according to (5.71) into

$$\mathbf{B} = \left[-\tfrac{3}{2L} \;\; \tfrac{2}{L} \;\; -\tfrac{1}{2L}\right] + \left[\tfrac{1}{L^2} \;\; -\tfrac{2}{L^2} \;\; \tfrac{1}{L^2}\right] x \tag{5.76}$$

The individual contributions to the element stiffness matrix are determined by the matrices

$$\mathbf{K}_0 = \frac{1}{L^2}\begin{bmatrix} \frac{9}{4} & -3 & \frac{3}{4} \\ -3 & 4 & -1 \\ \frac{3}{4} & -1 & \frac{1}{4} \end{bmatrix}; \; \mathbf{K}_1 = \frac{1}{L^3}\begin{bmatrix} -3 & 5 & -2 \\ 5 & -8 & 3 \\ -2 & 3 & -1 \end{bmatrix}; \; \mathbf{K}_2 = \frac{1}{L^4}\begin{bmatrix} 1 & -2 & 1 \\ -2 & 4 & -2 \\ 1 & -2 & 1 \end{bmatrix} \tag{5.77}$$

so that

$$\mathbf{K}^e = \mathbf{K}_0 X_0 + \mathbf{K}_1 X_1 + \mathbf{K}_2 X_2 \tag{5.78}$$

in which the weighted integrals X_k are defined by

$$X_k = A \int_0^{2L} E(x)x^k \, \mathrm{d}x \tag{5.79}$$

Using this definition, the mean values of the weighted integrals are easily computed as

$$\bar{X}_k = \mathbf{E}\left[A \int_0^{2L} E(x)x^k \, \mathrm{d}x\right] = A \int_0^{2L} \mathbf{E}[E(x)]x^k \, \mathrm{d}x = \bar{E}A\frac{(2L)^{k+1}}{k+1} \tag{5.80}$$

and the covariance C_{ik} between two weighted integrals is computed from

$$C_{ik} = A^2 \int_0^{2L}\int_0^{2L} C_{EE}(x,y)x^i y^k \, \mathrm{d}x\mathrm{d}y \tag{5.81}$$

Using `maxima` these integrals are evaluated in closed form as

$$\begin{aligned}
C_{00} &= \sigma_E^2 \cdot (4\sqrt{2}\arctan\sqrt{2} - \log 9)L^2A^2 = 3.20686L^2A^2\sigma_E^2 \\
C_{10} &= C_{01} = \sigma_E^2 \cdot (4\sqrt{2}\arctan\sqrt{2} - \log 9)L^3A^2 = 3.20686L^3\sigma_E^2A^2 \\
C_{20} &= C_{02} = \sigma_E^2 \cdot \frac{4\sqrt{2}\arctan\sqrt{2} - \log 6561 + 16}{3}L^4A^2 = 4.20506L^4\sigma_E^2A^2 \\
C_{11} &= \sigma_E^2 \cdot \frac{16\sqrt{2}\arctan\sqrt{2} - \log 4782969 + 4}{3}L^4A^2 = 3.41192L^4A^2\sigma_E^2 \\
C_{21} &= C_{12} = \sigma_E^2 \cdot (4\sqrt{2}\arctan\sqrt{2} - \log 6561 + 8)L^5A^2 = 4.61519L^5A^2\sigma_E^2 \\
C_{22} &= \sigma_E^2 \cdot \Big(\frac{16\sqrt{2}\arctan\sqrt{2} + 88\log 12 - 210\log 6 + 2\log 2}{15} \\
&\quad + \frac{16\sqrt{2}\arctan\sqrt{2} - 122\log 6 - 88\log 4 + 330\log 2 + 320}{15}\Big)L^6A^2 \\
&= 6.34475L^6A^2\sigma_E^2
\end{aligned} \tag{5.82}$$

so that the covariance matrix of the 3 weighted integrals becomes

$$\mathbf{C} = \sigma_E^2 \cdot \begin{bmatrix} 3.20686\,L^2 & 3.20686\,L^3 & 4.20506\,L^4 \\ 3.20686\,L^3 & 3.41192\,L^4 & 4.61519\,L^5 \\ 4.20506\,L^4 & 4.61519\,L^5 & 6.34475\,L^6 \end{bmatrix} A^2 \tag{5.83}$$

The Cholesky factor of the covariance matrix of the weighted integrals becomes

$$\mathbf{L} = \sigma_E \cdot \begin{bmatrix} 1.79077\,L & 0 & 0 \\ 1.79077\,L^2 & 0.45284\,L^2 & 0 \\ 2.34818\,L^3 & 0.90569\,L^3 & 0.10247\,L^3 \end{bmatrix} A \tag{5.84}$$

The `maxima` code for C_{21} is given in Listing 5.1.

```
1 c:(1/(1+(x-y)^2/2/L^2))*x*y^2;
2 d:integrate(c,x,0,2*L);
3 e:integrate(d,y,0,2*L);
4 f:logcontract(e);
5 bfloat(f);
```

Listing 5.1 Computation of covariance for weighted integrals

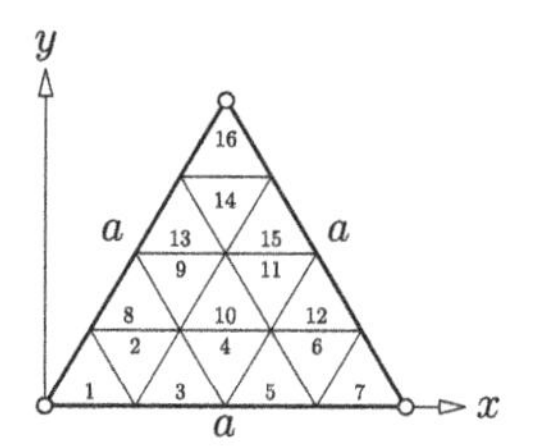

Figure 5.22 CST element subdivided into 16 subtriangles.

Example 5.10 (Weighted integral for CST element)

Consider the triangular element as discussed in Exercise 5.1. Compute the variance of the only weighted integral for this element if the covariance function of the modulus of elasticity is given by

$$C_{EE}(\mathbf{x}, \mathbf{y}) = \sigma_E^2 \exp\left(-\frac{||\mathbf{x} - \mathbf{y}||^2}{0.25L^2}\right) \tag{5.85}$$

Note: $||.||$ denotes Euclidian norm.

We choose a numerical scheme in which we subdivide the triangle in 16 smaller triangles. The integrand is represented by its values in the center points of the triangles.

The `octave` script is given below

```
1 function cee=cee(x,y)
2 dist=(x(1)-y(1))^2+(x(2)-y(2))^2;
3 cee=exp(-dist/.25);
4 endfunction
5 #
6 a=1;
7 b=a/8.
8 h=a*sqrt(3)/2/12
9 x1=[b,2*b, 3*b,4*b,5*b,6*b,7*b,\
10 2*b,3*b,4*b,5*b,6*b,\
11 3*b,4*b,5*b,\
12 4*b]
13 x2=[h,2*h,h,2*h,h,2*h,h,\
14 4*h,5*h,4*h,5*h,4*h,\
```

```
15 7*h,8*h,7*h,\
16 10*h]
17 tr1=[0;1;.5;0];
18 tr2=[0;0;sqrt(3)/2;0];
19 plot(x1, x2, "k+", tr1, tr2, "k-");
20 axis square
21 print('cstweight.eps','-deps')
22 cxx=0;
23 for i=1:16
24 for k=1:16
25   x=[x1(i),x2(i)];
26   y=[x1(k),x2(k)];
27   cxx+=cee(x,y)/256;
28 endfor
29 endfor
30 cxx
31 pause;
```

with the result

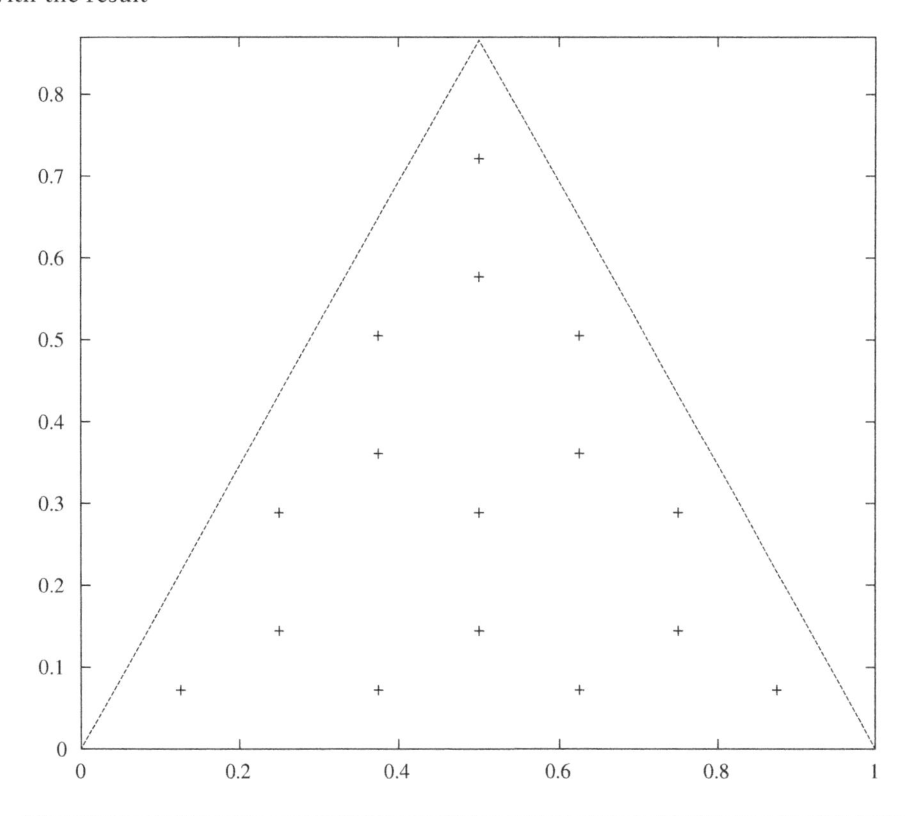


```
1 cxx = 0.60607
```

5.3.6 *Integration point method*

It turns out that the weighted integral representation, although rather elegant, is not particularly suitable for numerical solution procedures. The reasons may by summarized as follows:

- In general the element integrations are carried out numerically using Gaussian integration. This means that the number of integration points is smaller than the order of the polynomials to be integrated. As a consequence, there are more weighted integrals than there are random variables. This implies that the covariance matrix of the weighted integrals is singular (actually, the rank defect can be rather large).
- For non-Gaussian random fields it is almost impossible to derive the probability density functions of the weighted integrals. As a consequence it will not be possible to derive a full probabilistic description of the response.

As an alternative, the random field is represented pointwise, and in the context of Finite Element analysis, these points are optimally chosen to be the integration points of the element matrices. This leads to a representation of the element stiffness matrix in the form (Brenner 1995; Brenner and Bucher 1995)

$$\mathbf{K}^e = \sum_{g=1}^{N_g} \mathbf{K}_g^e E(\mathbf{x}_g^e) \tag{5.86}$$

Here $\mathbf{x}_g^e$ denotes the integration points in element e. All integration weights are included in the matrices $\mathbf{K}_g^e$. Then the global stiffness matrix is

$$\mathbf{K} = \bigoplus_{e=1}^{N_e} \sum_{g=1}^{N_g} \mathbf{K}_g^e E(\mathbf{x}_g^e) \tag{5.87}$$

In this representation, only the values of discretized random field are needed.

Example 5.11 (Rod Element) We discuss the three-node rod element mentioned before. Now we want to apply Gauss integration to obtain the element stiffness matrix. In order to integrate a polynomial of order $2m-1$ exactly, we need m integration points. So for the second order polynomials of the $\mathbf{B}\mathbf{B}^T$ matrix we need 2 integration points. In this way, we represent the integrals as

$$\int_0^{2L} p(x)\mathrm{d}x = 2L[w_1 p(x_1) + w_2 p(x_2)] \tag{5.88}$$

in which $w_1 = w_2 = \frac{1}{2}$ and $x_1 = L(1 - \frac{1}{\sqrt{3}})$, $x_2 = L(1 + \frac{1}{\sqrt{3}})$. In this representation, the element stiffness matrix becomes

$$\mathbf{K}^e = \left(\frac{E(x_1)}{2} \mathbf{B}(x_1)\mathbf{B}(x_1)^T + \frac{E(x_2)}{2} \mathbf{B}(x_2)\mathbf{B}(x_2)^T \right) 2LA \tag{5.89}$$

in which (cf (5.60))

$$\mathbf{B}(x_1) = \left[-\frac{\sqrt{3}+2}{2\sqrt{3}L} \; \frac{2}{\sqrt{3}L} \; \frac{\sqrt{3}-2}{2\sqrt{3}L} \right]; \quad \mathbf{B}(x_2) = \left[-\frac{\sqrt{3}-2}{2\sqrt{3}L} \; -\frac{2}{\sqrt{3}L} \; \frac{\sqrt{3}+2}{2\sqrt{3}L} \right] \tag{5.90}$$

so that

$$\mathbf{K}^e = \frac{E(x_1)A}{12L} \cdot \begin{bmatrix} 4\sqrt{3}+7 & -4\sqrt{3}-8 & 1 \\ -4\sqrt{3}-8 & 16 & 4\sqrt{3}-8 \\ 1 & 4\sqrt{3}-8 & 7-4\sqrt{3} \end{bmatrix} + \frac{E(x_2)A}{12L} \cdot \begin{bmatrix} 7-4\sqrt{3} & 4\sqrt{3}-8 & 1 \\ 4\sqrt{3}-8 & 16 & -4\sqrt{3}-8 \\ 1 & -4\sqrt{3}-8 & 4\sqrt{3}+7 \end{bmatrix} \tag{5.91}$$

5.3.7 *Static response – perturbation method*

Based on a perturbation approach, e.g. for random elastic modulus in the form

$$E(x,y) = E_0(x,y) + \epsilon E_1(x,y) \tag{5.92}$$

in which E_0 is the deterministic mean, ϵ is a "small" quantity, and E_1 describes the random deviation from the mean, the assembly of element stiffness matrices leads to a global stiffness matrix $\mathbf{K}$ which has random perturbations as well

$$\mathbf{K} = \mathbf{K}_0 + \epsilon \mathbf{K}_1 \tag{5.93}$$

Given the (deterministic) global load vector $\mathbf{P}$, the (random) global displacement vector $\mathbf{U}$ is determined from the solution of the following system of linear equations

$$\mathbf{KU} = \mathbf{P} \tag{5.94}$$

Expanding $\mathbf{U}$ into a power series with respect to ϵ

$$\mathbf{U} = \mathbf{U}_0 + \epsilon \mathbf{U}_1 + \epsilon^2 \mathbf{U}_2 \ldots \tag{5.95}$$

an approximate solution can be obtained in terms of powers of ϵ

$$\begin{aligned} \epsilon^0 &: \mathbf{K}_0\mathbf{U}_0 = \mathbf{F} \\ \epsilon^1 &: \mathbf{K}_0\mathbf{U}_1 = -\mathbf{K}_1\mathbf{U}_0 \\ \epsilon^2 &: \mathbf{K}_0\mathbf{U}_2 = -\mathbf{K}_1\mathbf{U}_1 \\ \epsilon^3 &: \ldots \end{aligned} \tag{5.96}$$

Usually, this is truncated after the linear terms in ϵ. From this first order perturbation result $\mathbf{U} = \mathbf{U}_0 + \epsilon \mathbf{U}_1$ the mean value $\bar{\mathbf{U}}$ of the displacement vector becomes

$$\bar{\mathbf{U}} = \mathbf{U}_0 \tag{5.97}$$

and the zero-mean random perturbation of the displacement vector is given by

$$\epsilon \mathbf{U}_1 = -\epsilon \mathbf{K}_0^{-1} \mathbf{K}_1 \mathbf{U}_0 \tag{5.98}$$

From this, the covariance matrix of its components is

$$C_{UU} = \epsilon^2 \mathbf{E}[U_1 U_1^T] = \epsilon^2 \mathbf{E}[K_0^{-1} K_1 U_0 U_0^T K_1^T K_0^{-1T}] = \epsilon^2 K_0^{-1} \mathbf{E}[K_1 U_0 U_0^T K_1^T] K_0^{-1T} \quad (5.99)$$

The actual computation can be based on the weighted integral representation (cf. Eq. 5.74). Let the global stiffness matrix be represented by

$$\mathbf{K}_1 = \bigoplus_{e=1}^{N_e} \sum_{k+l\leq r} \sum_{m+n\leq r} \mathbf{B}_{kl}^T \mathbf{D}_0 \mathbf{B}_{mn} X_{klmn}^e \quad (5.100)$$

Here, the first summation symbol indicates appropriate placement of the element stiffness matrices in the global stiffness matrix. After rearranging all weighted integrals into a vector **X** this can be rewritten as

$$\mathbf{K}_1 = \sum_{w=1}^{N_w} \mathbf{K}_w X_w \quad (5.101)$$

in which the matrices $\mathbf{K}_w$ are deterministic. The product of $\mathbf{K}_1$ and $\mathbf{U}_0$ becomes

$$\mathbf{K}_1 \mathbf{U}_0 = \sum_{w=1}^{N_w} \mathbf{K}_w \mathbf{U}_0 X_w = \sum_{w=1}^{N_w} \mathbf{r}_w X_w \quad (5.102)$$

Assembling all vectors $\mathbf{r}_w$ into a matrix **R** and all weighted integrals into a vector **X**, we obtain

$$\mathbf{K}_1 \mathbf{U}_0 = \mathbf{RX} \quad (5.103)$$

so that the covariance matrix of the displacement response becomes

$$C_{UU} = \epsilon^2 \mathbf{K}_0^{-1} \mathbf{R} \mathbf{E}[\mathbf{X}\mathbf{X}^T] \mathbf{R}^T \mathbf{K}_0^{-1T} \quad (5.104)$$

Using a second order expansion we can compute the additional perturbation term

$$\epsilon^2 \mathbf{U}_2 = -\epsilon \mathbf{K}_0^{-1} \mathbf{K}_1 \mathbf{U}_1 = \epsilon^2 \mathbf{K}_0^{-1} \mathbf{K}_1 \mathbf{K}_0^{-1} \mathbf{K}_1 \mathbf{U}_0 \quad (5.105)$$

It can be seen that the expected value $\mathbf{E}[U]$ then contains an additional term so that we get

$$\mathbf{E}[U] = \left(\mathbf{I} + \epsilon^2 \mathbf{E}[\mathbf{K}_0^{-1} \mathbf{K}_1 \mathbf{K}_0^{-1} \mathbf{K}_1]\right) \mathbf{U}_0 \quad (5.106)$$

Example 5.12 (Cantilever Rod)
A rod is modeled by two two-node elements as shown in Fig. 5.23. We assume that the modulus of elasticity $E(x)$ is a homogeneous random field with mean value $\bar{E}$, a coefficient of variation of 0.1 and a covariance function of $C_{EE}(x,y) = \sigma_E^2 \frac{1}{1+\frac{(x-y)^2}{2L^2}}$. We want to compute the mean values and the covariance matrix of the nodal displacements

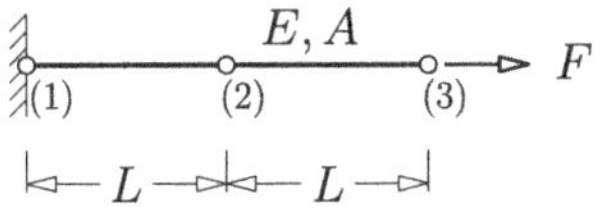

Figure 5.23 Cantilever rod with two elements.

using the weighted integral approach. The element stiffness matrices are easily found to be

$$\mathbf{K}^{(1)} = \frac{A}{L^2}\begin{bmatrix} 1 & -1 \\ -1 & 1 \end{bmatrix}\int_0^L E(x)\mathrm{d}x; \ \mathbf{K}^{(2)} = \frac{A}{L^2}\begin{bmatrix} 1 & -1 \\ -1 & 1 \end{bmatrix}\int_L^{2L} E(x)\mathrm{d}x \tag{5.107}$$

Defining the weighted integrals by

$$X_1 = \int_0^L (E(x) - \bar{E})\mathrm{d}x, \quad X_2 = \int_L^{2L} (E(x) - \bar{E})\mathrm{d}x \tag{5.108}$$

we can write the global stiffness matrix of the unsupported structure as

$$\mathbf{K}^u = \frac{\bar{E}A}{L}\begin{bmatrix} 1 & -1 & 0 \\ -1 & 2 & -1 \\ 0 & -1 & 1 \end{bmatrix} + \frac{A}{L^2}\begin{bmatrix} 1 & -1 & 0 \\ -1 & 1 & 0 \\ 0 & 0 & 0 \end{bmatrix} X_1 + \frac{A}{L^2}\begin{bmatrix} 0 & 0 & 0 \\ 0 & 1 & -1 \\ 0 & -1 & 1 \end{bmatrix} X_2 \tag{5.109}$$

Due to the boundary condition $U_1 = 0$, the first row and column of $\mathbf{K}$ has to be eliminated, so that

$$\mathbf{K} = \frac{\bar{E}A}{L}\begin{bmatrix} 2 & -1 \\ -1 & 1 \end{bmatrix} + \frac{A}{L^2}\begin{bmatrix} 1 & 0 \\ 0 & 0 \end{bmatrix} X_1 + \frac{A}{L^2}\begin{bmatrix} 1 & -1 \\ -1 & 1 \end{bmatrix} X_2 \tag{5.110}$$

The load vector $\mathbf{P}$ is given by

$$\mathbf{P} = \begin{bmatrix} 0 \\ f \end{bmatrix} \tag{5.111}$$

So we have

$$\mathbf{U}_0 = \left(\frac{\bar{E}A}{L}\begin{bmatrix} 2 & -1 \\ -1 & 1 \end{bmatrix}\right)^{-1}\begin{bmatrix} 0 \\ f \end{bmatrix} = \frac{L}{\bar{E}A}\begin{bmatrix} 1 & 1 \\ 1 & 2 \end{bmatrix}\begin{bmatrix} 0 \\ f \end{bmatrix} = \frac{fL}{\bar{E}A}\begin{bmatrix} 1 \\ 2 \end{bmatrix} \tag{5.112}$$

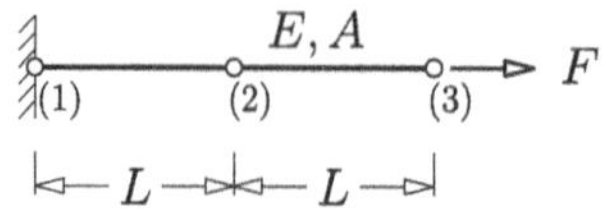

E, A
(1) (2) (3) F
L L

Figure 5.24 Cantilever rod with one three-node element.

The covariance matrix of the weighed integrals is computed using `maxima` as

$$\mathbf{C}_{XX} = \sigma_E^2 L^2 \begin{bmatrix} 2\sqrt{2}\arctan\frac{\sqrt{2}}{2} + \log\frac{4}{9} & 2\sqrt{2}(\arctan\sqrt{2} - \arctan\frac{\sqrt{2}}{2}) + \log\frac{3}{4} \\ \text{sym.} & 2\sqrt{2}\arctan\frac{\sqrt{2}}{2} + \log\frac{4}{9} \end{bmatrix}$$

$$= \sigma_E^2 L^2 \begin{bmatrix} 0.92991 & 0.67352 \\ 0.67352 & 0.92991 \end{bmatrix}$$

Inserting these results into (5.104) we finally obtain

$$\mathbf{C}_{UU} = \left(\frac{fL}{\bar{E}A}\right)^2 \begin{bmatrix} 0.0092991 & 0.0160343 \\ 0.0160343 & 0.0320686 \end{bmatrix}$$

which leads to coefficients of variation for U_2 and U_3

$$COV_2 = \frac{\sqrt{0.0092991}}{1} = 0.0964; \quad COV_3 = \frac{\sqrt{0.0320686}}{2} = 0.0895$$

5.3.8 *Monte Carlo simulation*

The integration point method is easily incorporated into a Monte Carlo method. In this process, it is required to apply a suitable joint probability density model for the random variables $\mathbf{X}$ representing the discretized random field. In most situations, the Nataf-model as described in section 2.2.6 is appropriate. In order to apply the Nataf-model, the modified coefficients of correlation in the space of Gaussian variables $\mathbf{V}$ must be computed using Eq. 2.82. Based on this, the covariance matrix $\mathbf{C}_{VV}$ can be formed and Cholesky-decomposed. Finally, samples of uncorrelated standardized Gaussian variables $\mathbf{U}$ are generated, transformed to the space $\mathbf{V}$, and finally using the inverse relation of (2.78)

$$X_i = F_{X_i}^{-1}[\Phi(V_i)] \tag{5.113}$$

transformed back to physical space. Once samples of the discretized random field have been generated, the element matrices are assembled in straightforward manner according to (5.86), and solutions for the displacements can be computed.

Example 5.13 (Cantilever Rod)

Consider a beam modeled by one three-node rod element as shown in Fig. 5.24. The elastic modulus is assumed to be a log-normally distributed random field with a mean value of $\bar{E}$, a coefficient of variation of 0.2, and a covariance function $C_{EE}(x,y) = \sigma_{\bar{E}}^2 \frac{1}{1+\frac{(x-y)^2}{2L^2}}$. We want to compute the mean and standard deviation of the end displacement. The global stiffness matrix is easily obtained from the previous example by eliminating the first row and column out of (5.91)

$$\mathbf{K} = \frac{E(x_1)A}{12L} \cdot \begin{bmatrix} 16 & 4\sqrt{3}-8 \\ 4\sqrt{3}-8 & 7-4\sqrt{3} \end{bmatrix} + \frac{E(x_2)A}{12L} \cdot \begin{bmatrix} 16 & -4\sqrt{3}-8 \\ -4\sqrt{3}-8 & 4\sqrt{3}+7 \end{bmatrix}$$

The distance between the integration points is $x_2 - x_1 = \frac{2L}{\sqrt{3}}$ so that the covariance between $E(x_1)$ and $E(x_2)$ becomes $\sigma_{EE}^2 \frac{1}{1+\frac{4}{3 \cdot 2}} = 0.6\sigma_{EE}^2$. So the coefficient of correlation between the two random variables is $\rho_{X_1X_2} = 0.6$. The modified coefficients of correlation in the space of Gaussian variables V_1, V_2 as required by the Nataf-model (cf. section 2.2.6) are computed as $\rho_{V_1V_2} = 0.606$. Based on (2.35) and (2.36) the transformation from **V** to **X** space becomes

$$X_k = \mu \exp[s \cdot V_k]; \quad k = 1, 2$$

in which $s = 0.198$ and $\mu = 0.9806\bar{E}$.

The octave-script for the solution of the problem is given in Listing 5.2

```
1  k1=[16,4*sqrt(3)-8;4*sqrt(3)-8,7-4*sqrt(3)]/12;
2  k2=[16,-4*sqrt(3)-8;-4*sqrt(3)-8,4*sqrt(3)+7]/12;
3  cvv=[1,.606;.606,1];
4  l=chol(cvv)';
5  mu=0.9806;
6  s=0.198;
7  N=10000
8  del=[];
9  f=[0;1];
10 for i=1:N
11   u=randn(2,1);
12   v=l*u;
13   x=mu*exp(s*v);
14   k=x(1)*k1+x(2)*k2;
15   disp=inv(k)*f;
16   del=[del;disp(2)];
17 endfor
18 dm=mean(del)
19 ds=std(del)
20 cov=ds/dm
```

Listing 5.2 Computation of statistics of cantilever rod displacement

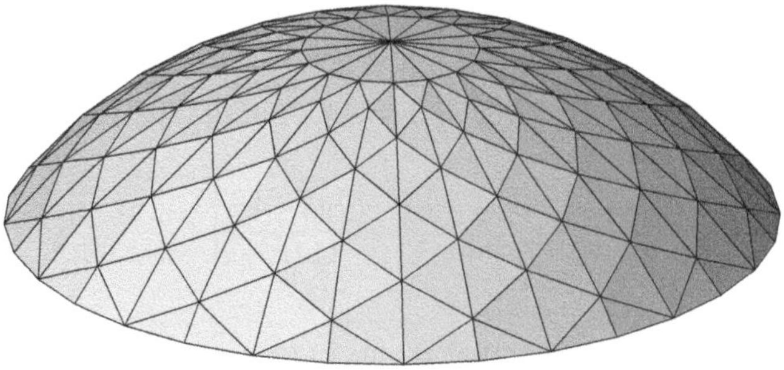

Figure 5.25 Spherical Shell Structure.

The result is

```
1 N = 10000
2 dm = 2.0843
3 ds = 0.37264
4 cov = 0.17878
```

which means that the mean value is computed as $\bar{U}_3 = \frac{2.08L}{\bar{E}A}$ and the coefficient of variation is 0.18, slightly smaller than the coefficient of variation of the modulus of elasticity.

5.3.9 *Natural frequencies of a structure with randomly distributed elastic modulus*

A spherical shell structure (cf. Fig. 5.25) in free vibration is considered. The shell is modeled by 288 triangular elements of the type SHELL3N (Schorling 1997). The shell is assumed to be simply supported along the edge. The material of the shell is assumed to be elastic (in plain stress) and the elastic modulus $E(\mathbf{x})$ is modeled as a log-normally distributed, homogeneous, and isotropic random field. Its auto-covariance function $C_{EE}(\mathbf{x}, \mathbf{y})$ is assumed to be of the form

$$C_{EE}(\mathbf{x}, \mathbf{y}) = \sigma_{EE}^2 \exp\left(-\frac{||\mathbf{x} - \mathbf{y}||}{L_c}\right) \tag{5.114}$$

In the above equations, the 3D vectors $\mathbf{x}$ and $\mathbf{y}$ are the coordinates of points within the shell structure. The basis diameter of the shell is 20 m, and the correlation length L_c is assumed to be 10 m. The coefficient of variation of the random field is assumed to be 0.2.

The question to be answered is what magnitude of randomness may be expected in the calculated natural frequencies? Quite clearly, this is very important for structural elements designed to carry, e.g., rotating machinery which produces almost harmonic excitation, and possibly resonance. Hence the probability of obtaining high deviations from the mean natural frequency needs to be calculated. This example shows, quite

Figure 5.26 Selected mode shapes of the covariance function.

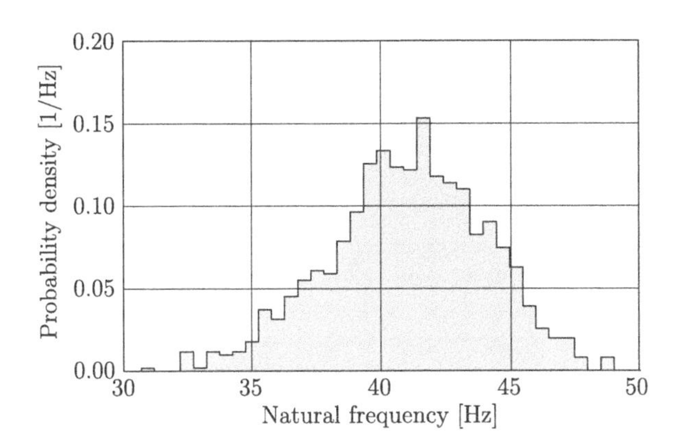

Figure 5.27 Histogram of the fundamental natural frequency.

typically, the close connection which is required between the stochastic analyses and the finite element analysis. Within the finite element model, the random field $E(\mathbf{x})$ is represented by its values in the integration points of the elements. The shell elements as utilized here have two layers of each 13 integration points, so there is a total of $26 \times 288 = 7488$ integration points. In order to reduce this rather high number of random variables, the following strategy is applied. First, the elastic modulus is represented by one value per element (given the correlation length of 10 m this is not a severe simplification). Second, the remaining random field is represented in terms of independent random variables and corresponding space dependent shape functions. These independent variables are obtained by applying the Nataf joint density model along with an eigenvalue decomposition of the covariance matrix of the random field, i.e. by the spectral representations outlined earlier in this chapter.

Fig. 5.26 shows selected space dependent shape functions. They are ordered according to decreasing magnitude of the corresponding eigenvalues. A Monte Carlo simulation is then carried out to generate sample functions of the random field. For each sample, the lowest natural frequency is computed. The simulation results obtained from 1000 samples in terms of histograms are given in Fig. 5.27.

The results indicate a relatively large scatter of the fundamental frequency (the deterministic system has two identical lowest natural frequencies at 42 Hz). The coefficient of variation is approximately 11%.

Summary

This chapter introduced the definitions and basic properties of random fields. The spectral representation of random fields was discussed in the continuous and discrete cases. Conditional random fields were presented in order to introduce deterministic values at predefined locations. Properties of local averages of random fields were described. The random field modeling of geometrical imperfections was introduced. The basic concepts of stochastic finite elements were derived from the principle of virtual work and standard finite element technology. Expressions for the structural response were derived in terms of weighted integrals of the random fields. As an alternative, perturbation methods were discussed. Finally, Monte Carlo methods were presented and applied.

Chapter 6

Computation of failure probabilities

ABSTRACT: Structural safety requires the design of structures in a way that the probability of failure becomes extremely small. Although the computation of the failure probability basically only requires numerical integration, standard computational techniques are not capable of handling this task efficiently. Hence, methods were developed which aim at providing better computational tools for reliability analysis. A representative selection of these methods is presented in this chapter together with numerical applications.

6.1 Structural reliability

6.1.1 *Definitions*

Generally, failure (i.e. an undesired or unsafe state of the structure) is defined in terms of a limit state function $g(.)$, i.e. by the set $\mathcal{F} = \{\mathbf{X} : g(\mathbf{X}) \leq 0\}$. Frequently, $Z = g(\mathbf{X})$ is called *safety margin*.

For the simple problem as shown in Fig. 6.1, the definition of the limit state function is not unique, i.e. there are several ways of expressing the failure condition

$$\mathcal{F} = \{(F, L, M_{pl}) : FL \geq M_{pl}\} = \left\{(F, L, M_{pl}) : 1 - \frac{FL}{M_{pl}} \leq 0\right\} \tag{6.1}$$

The failure probability is defined as the probability of the occurrence of $\mathcal{F}$:

$$p_f = \textbf{Prob}[\{\mathbf{X} : g(\mathbf{X}) \leq 0\}] \tag{6.2}$$

This quantity is *unique*, i.e. not depending on the particular choice of the limit state function.

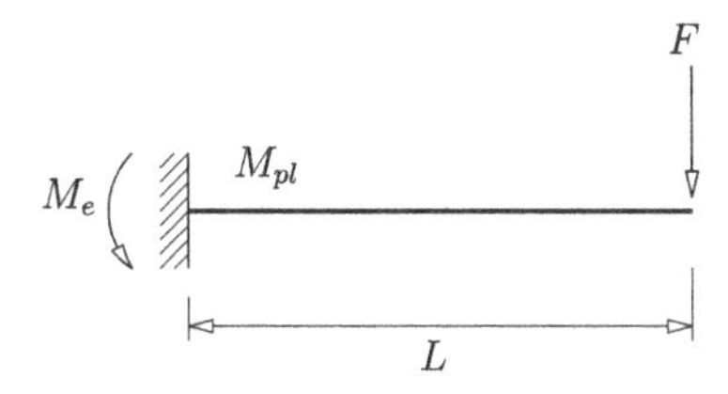

Figure 6.1 Simple structural system.

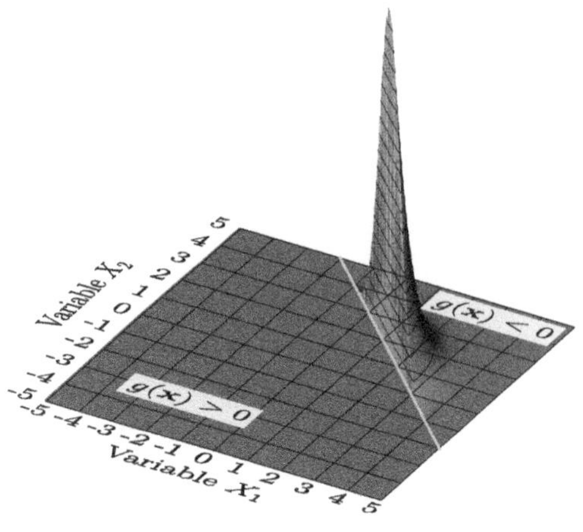

Figure 6.2 Integrand for calculating the probability of failure for $g(x_1, x_2) = 3 - x_1 - x_2$.

The failure probability can be written in the form of a multi-dimensional integral

$$p_f = \mathbf{Prob}[g(X_1, X_2, \ldots X_n) \leq 0] = \int \cdots \int_{g(\mathbf{X}) \leq 0} f_{\mathbf{X}}(\mathbf{x}) d\mathbf{x} \tag{6.3}$$

The computational challenge in determining the integral of Eq. 6.3 lies in evaluating the limit state function $g(\mathbf{x})$, which for non-linear systems usually requires an incremental/iterative numerical approach. In this context, it is essential to realize that the limit state function $g(\mathbf{x})$ serves the sole purpose of defining the bounds of integration in Eq. (6.3). As an example, consider a 2-dimensional problem with standard normal random variables X_1 and X_2, and a limit state function $g(x_1, x_2) = 3 - x_1 + x_2$. In Fig. 6.2 the integrand of Eq. 6.3 in the failure domain is displayed. It is clearly visible that only a very narrow region around the so-called design point $\mathbf{x}^*$ really contributes to the value of the integral, i.e., the probability of failure $P(\mathcal{F})$. This makes is difficult to locate integration points for numerical integration procedures appropriately.

6.1.2 *First order – second moment concept*

The first-order second moment method aims at a representation of the limit state function $g(.)$ by a Taylor series and subsequent calculation of the statistical moments of the safety margin Z.

$$g(\mathbf{x}) = g(\mathbf{x}_0) + \sum_{i=1}^{n} \left.\frac{\partial g}{\partial x_i}\right|_{\mathbf{x}=\mathbf{x}_0} (x_i - x_{i0})$$
$$+ \frac{1}{2} \sum_{i=1}^{n} \sum_{k=1}^{n} \left.\frac{\partial^2 g}{\partial x_i \partial x_k}\right|_{\mathbf{x}=\mathbf{x}_0} (x_i - x_{i0})(x_k - x_{k0}) + \ldots \tag{6.4}$$

Terminating the series after the *quadratic* terms yields

$$\mathbf{E}[Z] = \mathbf{E}[g(\mathbf{X})] = g(\mathbf{x}_0) + \sum_{i=1}^{n} \frac{\partial g}{\partial x_i} \mathbf{E}[x_i - x_{i0}]$$
$$+ \frac{1}{2} \sum_{i=1}^{n} \sum_{k=1}^{n} \frac{\partial^2 g}{\partial x_i \partial x_k} \mathbf{E}[(x_i - x_{i0})(x_k - x_{k0})] \tag{6.5}$$

Terminating the Taylor series after the *linear* terms yields

$$\mathbf{E}[Z] = \mathbf{E}[g(\mathbf{X})] = g(\mathbf{x}_0) + \sum_{i=1}^{n} \frac{\partial g}{\partial x_i} \mathbf{E}[x_i - x_{i0}] \tag{6.6}$$

If the mean value vector $\bar{\mathbf{X}}$ is chosen as expansion point $\mathbf{x}_0$ for the Taylor series, then $E[Z] = g(\mathbf{x}_0)$ and the variance becomes

$$\sigma_Z^2 = \mathbf{E}[(Z - \bar{Z})^2] = \mathbf{E}\left[\left(\sum_{i=1}^{n} \frac{\partial g}{\partial x_i}(X_i - \bar{X}_i)\right)^2\right]$$
$$= \sum_{i=1}^{n} \sum_{k=1}^{n} \frac{\partial g}{\partial x_i} \frac{\partial g}{\partial x_k} \mathbf{E}[(X_i - \bar{X}_i)(X_k - \bar{X}_k)] \tag{6.7}$$

Finally the distribution function $F_Z(z)$ is approximated by a normal distribution

$$F_Z(z) = \Phi\left(\frac{z - \bar{Z}}{\sigma_Z}\right) \tag{6.8}$$

Then we obtain the approximate result

$$p_f = F_Z(0) = \Phi\left(-\frac{\bar{Z}}{\sigma_Z}\right) \tag{6.9}$$

Note that this result does not take into account the types of distributions of the basic variables. It also depends significantly on the choice of the expansion point for the Taylor-series.

Example 6.1 (FOSM – quadratic problem)
Consider a reliability problem involving two uncorrelated random variables X_1 and X_2. Assume that both variables have mean values of 1 and standard deviations of 0.5. The limit state function is

$$g(x_1, x_2) = 4 - x_1^2 - x_2 \tag{6.10}$$

Linearizing about the mean values we get the first-order approximation

$$\hat{g}(x_1, x_2) = 2 - 2(x_1 - 1) - (x_2 - 1) \tag{6.11}$$

which leads to

$$\mathbf{E}[Z] = 2; \quad \sigma_Z^2 = 4 \cdot 0.25 + 0.25 = 1.25 \tag{6.12}$$

resulting in

$$p_f = \Phi\left(-\frac{2}{\sqrt{1.25}}\right) = \Phi(1.7889) = 0.0368 \tag{6.13}$$

Choosing a different expansion point, we obtain a different result. For example, we might choose an expansion point such that its limit state function value is zero, e.g. the point (2,2). In this case we get

$$\hat{g}(x_1, x_2) = -4(x_1 - 2) - (x_2 - 2) = 5 - 4(x_1 - 1) - (x_2 - 1) \tag{6.14}$$

which results in

$$\mathbf{E}[Z] = 5; \quad \sigma_Z^2 = 16 \cdot 0.25 + 0.25 = 4.25 \tag{6.15}$$

and thus

$$p_f = \Phi\left(-\frac{5}{\sqrt{4.25}}\right) = \Phi(2.4254) = 0.00765 \tag{6.16}$$

It can be seen that the choice of the expansion point significantly influences the resulting failure probability.

6.1.3 FORM – *first order reliability method*

The FORM-Concept (Hasofer and Lind 1974) is based on a description of the reliability problem in standard Gaussian space. Hence transformations from correlated non-Gaussian variables $\mathbf{X}$ to uncorrelated Gaussian variables $\mathbf{U}$ with zero mean and unit variance are required. This step is called Rosenblatt-transformation. Then a linearization in performed in $\mathbf{u}$-space. The expansion point $\mathbf{u}^*$ is chosen such as to maximize the pdf within the failure domain. Geometrically, this coincides with the point in the failure domain, having the minimum distance β from the origin. From a safety engineering point of view, the point $\mathbf{x}^*$ corresponding to $\mathbf{u}^*$ is called *design point*.

This concept is especially useful in conjunction with the Nataf-model for the joint pdf of $\mathbf{X}$. In this case the Rosenblatt-transformation consists of the following steps:

1. Transform from correlated non-Gaussian variables X_i to correlated Gaussian variables Y_i

   $$Y_i = \Phi^{-1}[F_{X_i}(X_i)]; \quad i = 1 \ldots n \tag{6.17}$$

   These transformations can be carried out independently. The covariance matrix $\mathbf{C}_{\mathbf{YY}}$ is calculated from $\mathbf{C}_{\mathbf{XX}}$ according to the rules of the Nataf-model (cf. section 2.2.6).
2. Transform from correlated Gaussian space to standard Gaussian space by means of

   $$\mathbf{U} = \mathbf{L}^{-1}\mathbf{Y} \tag{6.18}$$

in which $\mathbf{L}$ is calculated from the Cholesky-decomposition of $\mathbf{C_{YY}}$

$$\mathbf{C_{YY}} = \mathbf{L}\mathbf{L}^T \tag{6.19}$$

In total, this leads to a representation of the limit state function $g(.)$ in terms of the standardized Gaussian variables U_i

$$g(\mathbf{X}) = g(X_1, X_2, \ldots X_n) = g[X_1(U_1, \ldots U_n) \ldots X_n(U_1, \ldots U_n)] \tag{6.20}$$

with

$$X_i = F_{X_i}^{-1}\left[\Phi\left(\sum_{k=1}^{n} L_{ik} U_k\right)\right] \tag{6.21}$$

From the geometrical interpretation of the expansion point $\mathbf{u}^*$ in standard Gaussian space it becomes quite clear that the calculation of the design point can be reduced to an optimization problem

$$\mathbf{u}^* = \text{argmin}\left(\frac{1}{2}\mathbf{u}^T\mathbf{u}\right); \quad \text{subject to: } g[\mathbf{x}(\mathbf{u})] = 0 \tag{6.22}$$

This leads to the Lagrange-function

$$L = \frac{1}{2}\mathbf{u}^T\mathbf{u} + \lambda g(\mathbf{u}) \rightarrow \text{Min.} \tag{6.23}$$

Standard optimization procedures can be utilized to solve for the location of $\mathbf{u}^*$. One of the earliest methods is the so-called Rackwith-Fiessler algorithm (Rackwitz and Fiessler 1978). This algorithm is a simple version of the SQP optimization procedure. In this procedure, the objective function is replaced by a quadratic approximation and the constrain conditions are linearized. In view of Lagrangian as given by Eq. 6.23 this means that the objective function is unchanged whereas the constraint is replaced by the linearized version using $\mathbf{u} = \mathbf{u}_0 + \mathbf{v}$

$$\hat{g}(\mathbf{u}) = g(\mathbf{u}_0) + \nabla g(\mathbf{u}_0)^T(\mathbf{u} - \mathbf{u}_0) = g(\mathbf{u}_0) + \nabla g(\mathbf{u}_0)^T\mathbf{v} \tag{6.24}$$

In this equation, $\mathbf{u}_0$ is an expansion point, usually chosen to be the current iterate. The approximate Lagrangian

$$\begin{aligned}\hat{L} &= \frac{1}{2}\mathbf{u}^T\mathbf{u} + \lambda\hat{g}(\mathbf{u}) \\ &= \frac{1}{2}\mathbf{v}^T\mathbf{v} + \mathbf{u}_0^T\mathbf{v} + \frac{1}{2}\mathbf{u}_0^T\mathbf{u}_0 + \lambda\left[g(\mathbf{u}_0) + \nabla g(\mathbf{u}_0)^T\mathbf{v}\right]\end{aligned} \tag{6.25}$$

is associated with the Kuhn-Tucker conditions

$$\begin{aligned}&\mathbf{v} + \mathbf{u}_0 + \lambda\nabla g(\mathbf{u}_0) = 0 \\ &g(\mathbf{u}_0) + \nabla g(\mathbf{u}_0)^T\mathbf{v} = 0\end{aligned} \tag{6.26}$$

This system of equations is solved by

$$\lambda = \frac{g(\mathbf{u}_0) - \nabla g(\mathbf{u}_0)^T \mathbf{u}_0}{\nabla g(\mathbf{u}_0)^T \nabla g(\mathbf{u}_0)} \tag{6.27}$$

and

$$\mathbf{u}_0 + \mathbf{v} = -\lambda \nabla g(\mathbf{u}_0) \tag{6.28}$$

Then $\mathbf{u}_0$ is replaced by $\mathbf{u} = \mathbf{u}_0 + \mathbf{v}$ and the iteration proceeds from Eq. 6.24 until convergence of $\mathbf{u}_0$ to $\mathbf{u}^*$. It is known that this simple version of the algorithm does not always converge, hence more sophisticated optimization methods may be appropriate (e.g. NLPQL, Schittkowski 1986).

Once the point $\mathbf{u}^*$ is located, the exact limit state function $g(\mathbf{u})$ is replaced by a linear approximation $\hat{g}(\mathbf{u})$ as shown in Fig. 6.3. Geometrically, it can easily be seen that $\hat{g}(\mathbf{u})$ is determined from

$$\hat{g} : -\sum_{i=1}^{n} \frac{u_i}{s_i} + 1 = 0; \quad \sum_{i=1}^{n} \frac{1}{s_i^2} = \frac{1}{\beta^2} \tag{6.29}$$

The safety margin $Z = -\sum_{i=1}^{n} \frac{U_i}{s_i} + 1$ is normally distributed with the following statistical moments

$$E[Z] = 1; \quad \sigma_Z^2 = \sum_{i=1}^{n} \sum_{k=1}^{n} \frac{E[U_i U_k]}{s_i s_k} = \sum_{i=1}^{n} \frac{E[U_i^2]}{s_i^2} = \sum_{i=1}^{n} \frac{1}{s_i^2} = \frac{1}{\beta^2} \tag{6.30}$$

$$\rightarrow \sigma_Z = \frac{1}{\beta} \tag{6.31}$$

From this, the probability of failure is easily determined to be

$$p_f = \Phi\left(-\frac{1}{\frac{1}{\beta}}\right) = \Phi(-\beta) \tag{6.32}$$

This result is exact, if $g(\mathbf{u})$ is actually linear.

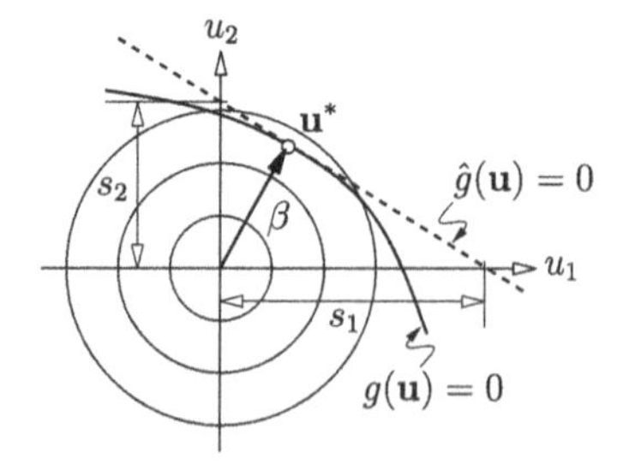

Figure 6.3 Linearization required for first order reliability method.

Example 6.2 (Application of FORM)
Consider a simple reliability problem with $g(X_1, X_2) = X_1 - X_2$ and let X_1 be Gaussian with $\bar{X}_1 = 1$, $\sigma_{X_1} = 0.4$ and X_2 be exponentially distributed with $\bar{X}_2 = \sigma_{X_2} = 0.2$. X_1 and X_2 are assumed to be stochastically independent. We want to calculate p_f based on FORM and compare the result to the exact solution (based on integration).

The unit random variables U_1 and U_2 can be obtained from

$$u_1 = \frac{x_1 - \bar{X}_1}{\sigma_{X_1}}; \quad u_2 = \Phi^{-1}\left[1 - \exp\left(-\frac{x_2}{\bar{X}_2}\right)\right] \tag{6.33}$$

The inverse relations are

$$x_1 = \sigma_{X_1} u_1 + \bar{X}_1; \quad x_2 = -\bar{X}_2 \log\left[1 - \Phi(u_2)\right] \tag{6.34}$$

The limit state function $g(.)$ is defined by

$$g(u_1, u_2) = \{\sigma_{X_1} u_1 + \bar{X}_1 + \bar{X}_2 \log\left[1 - \Phi(u_2)\right]\} \tag{6.35}$$

which defines the boundary between safe and failure domains by the condition $x_1 = x_2$, i.e. by

$$u_1 = -\frac{1}{\sigma_{X_1}} \{\bar{X}_1 + \bar{X}_2 \log\left[1 - \Phi(u_2)\right]\} \tag{6.36}$$

The gradient of the limit state function with respect to the variables u_1, u_2 in standard Gaussian space is given by

$$\nabla g(u_1, u_2) = \begin{bmatrix} \sigma_{X_1} \\ -\frac{\bar{X}_2}{\sqrt{2\pi}} \exp\left(-\frac{u_2^2}{2}\right) \frac{1}{1 - \Phi(u_2)} \end{bmatrix} \tag{6.37}$$

The Rackwitz-Fiessler iteration sequence starting from the origin in **u**-space is shown in Fig. 6.4. It can be clearly seen that convergence is very fast.

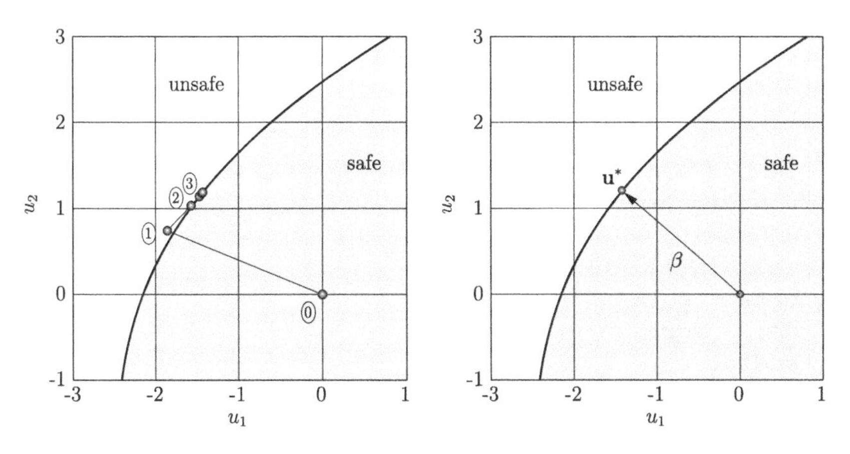

Figure 6.4 Rackwitz-Fiessler iteration sequence to design point.

The design point $\mathbf{u}^*$ has the coordinates $u_1^* = -1.4204$ and $u_2^* = 1.1982$. The safety index is found to be $\beta = 1.858$, and from that an approximation of the failure probability as $p_f = \Phi(-1.858) = 0.0316$. The exact failure probability (obtained by numerical integration) is $p_f = 0.0359$.

Transformation back into original space yields the design point $\mathbf{x}^*$ with the coordinates $x_1^* = x_2^* = 0.4318$ which, of course, satisfies the limit state condition in original space, i.e. $g(x_1^*, x_2^*) = x_1^* - x_2^* = 0$.

The octave-code for the Rackwitz-Fiessler iteration is given in Listing 6.1.

```
1  function grad=grad(u)
2    global s1;
3    global m1;
4    global m2;
5    grad=[s1;-m2/sqrt(2*pi)*exp(-u(2)^2/2)/(1-normcdf(u(2)))];
6    endfunction
7
8  function func=func(u)
9    global s1;
10   global m1;
11   global m2;
12   func = m1+s1*u(1) + m2*log(1-normcdf(u(2)))
13   endfunction
14
15 global s1 = 0.4;
16 global m1 = 1;
17 global m2 = 0.2;
18
19 u=[0;0];
20 uall=u';
21 for i=1:5
22   g = grad(u);
23   gtg = g'*g;
24   gtu = g'*u;
25   f = func(u);
26   lambda = (f-gtu)/gtg;
27   v = g*(-lambda) - u;
28   u = u+v;
29   uall=[uall;u'];
30   beta = sqrt(u'*u)
31   del = sqrt(v'*v);
32   endfor
33 save("-ascii","uall.txt","uall")
```

Listing 6.1 Rackwitz-Fiessler iteration.

6.2 Monte Carlo simulation

6.2.1 *Definitions and basics*

The definition of the failure probability as given in Eq. 6.3 can be written as an expected value

$$p_f = \int_{-\infty}^{\infty}\int_{-\infty}^{\infty}\cdots\int_{-\infty}^{\infty} I_g(x_1\ldots x_n) f_{X_1\ldots X_n}(x_1\ldots x_n)dx_1\ldots dx_n \tag{6.38}$$

in which $I_g(x_1\ldots x_n)=1$ if $g(x_1\ldots x_n)\leq 0$ and $I_g(.)=0$ else.

In order to determine p_f in principle all available statistical methods for estimation of expected values are applicable. If m independent samples $\mathbf{x}^{(k)}$ of the random vector $\mathbf{X}$ are available then the estimator

$$\bar{p}_f = \frac{1}{m}\sum_{k=1}^{m} I_g(\mathbf{x}^{(k)}) \tag{6.39}$$

yields a consistent and unbiased estimate for p_f.

The problem associated with this approach is this: For small values of p_f and small values of m the confidence of the estimate is very low. The variance $\sigma^2_{\bar{p}_f}$ of the estimate $\bar{p}_f$ can be determined from

$$\sigma^2_{\bar{p}_f} = \frac{p_f}{m} - \frac{p_f^2}{m} \approx \frac{p_f}{m} \rightarrow \sigma_{\bar{p}_f} = \sqrt{\frac{p_f}{m}} \tag{6.40}$$

It is to be noted that the required number m of simulations is independent of the dimension n of the problem.

6.2.2 *Importance sampling (weighted simulation)*

General concept

In order to reduce the standard deviation $\sigma_{\bar{p}_f}$ of the estimator to the order of magnitude of the probability of failure itself m must be in the range of $m=\frac{1}{p_f}$. For values of p_f in the range of 10^{-6} this cannot be achieved if each evaluation of the limit state function requires a complex structural analysis. Alternatively, strategies are employed which increase the "hit-rate" by artificially producing more samples in the failure domain than should occur according to the distribution functions. One way to approach this solution is the introduction of a positive weighting function $h_{\mathbf{Y}}(\mathbf{x})$ which can be interpreted as density function of a random vector $\mathbf{Y}$. Samples are taken according to $h_{\mathbf{Y}}(\mathbf{x})$.

The probability of failure is then estimated from

$$\bar{p}_f = \frac{1}{m}\sum_{k=1}^{m}\frac{f_{\mathbf{X}}(\mathbf{x})}{h_{\mathbf{Y}}(\mathbf{x})} I_g(\mathbf{x}) = \mathbf{E}\left[\frac{f_{\mathbf{X}}(\mathbf{x})}{h_{\mathbf{Y}}(\mathbf{x})} I_g(\mathbf{x})\right] \tag{6.41}$$

From the estimation procedure as outlined in section 2.3 it can be seen that the variance of the estimator $\bar{p}_f$ becomes

$$\sigma_{\bar{p}_f}^2 = \frac{1}{m}\mathbf{E}\left[\frac{f_{\mathbf{X}}(\mathbf{x})^2}{h_{\mathbf{Y}}(\mathbf{x})^2} I_g(\mathbf{x})\right] \tag{6.42}$$

A useful choice of $h_{\mathbf{Y}}(\mathbf{x})$ should be based on minimizing $\sigma_{\bar{p}_f}^2$. Ideally, the weighting function should reduce the sampling error to zero. However, this cannot be achieved in reality since such a function must have the property

$$h_{\mathbf{Y}}(\mathbf{x}) = \begin{cases} \frac{1}{p_f} f_{\mathbf{X}}(\mathbf{x}) & g(\mathbf{x}) \leq 0 \\ 0 & g(\mathbf{x}) > 0 \end{cases} \tag{6.43}$$

This property requires the knowledge of p_f which, of course, is unknown. Special updating procedures such as adaptive sampling (Bucher 1988a) can help to alleviate this problem.

Example 6.3 (1-Dimensional problem)
Let X be normally distributed with $f_X(x) = \frac{1}{\sqrt{2\pi}} \exp\left(-\frac{x^2}{2}\right)$. Assume that the limit state function is given by $g(x) = \beta - x$. We will try to find an optimal sampling density function in the form $h_Y(x) = \frac{1}{\sqrt{2\pi}} \exp\left(-\frac{(x-\bar{Y})^2}{2}\right)$. In this form, $\bar{Y}$ will be chosen to minimize the variance of the estimated failure probability.

This variance can be calculated directly by evaluating the expectations given above

$$\begin{aligned} \sigma_{\bar{p}_f}^2 &= \frac{1}{m}\int_\beta^\infty \frac{f_X(x)^2}{h_Y(x)^2} h_Y(x) \mathrm{d}x \\ &= \frac{1}{m}\int_\beta^\infty \frac{1}{\sqrt{2\pi}} \exp\left(-\frac{2x^2}{2} + \frac{(x-\bar{Y})^2}{2}\right)\mathrm{d}x \\ &= \frac{1}{m}\exp(\bar{Y}^2)\Phi[-(\beta+\bar{Y})] \end{aligned} \tag{6.44}$$

Differentiation with respect to $\bar{Y}$ yields

$$\begin{aligned} &\frac{\partial}{\partial \bar{Y}}(\sigma_{\bar{p}_f}^2) = 0 \\ &\rightarrow 2\bar{Y}\Phi[-(\beta+\bar{Y})] - \frac{1}{\sqrt{2\pi}}\exp\left(-\frac{(\beta+\bar{Y})^2}{2}\right) = 0 \end{aligned} \tag{6.45}$$

Using the following asymptotic (as $z \rightarrow \infty$) approximation for $\Phi(.)$ (Mill's ratio, see e.g. Abramowitz and Stegun 1970)

$$\Phi(-z) \approx \frac{1}{z\sqrt{2\pi}}\exp\left(-\frac{z^2}{2}\right) \tag{6.46}$$

an asymptotic solution to the minimization problem is given by

$$\frac{2\bar{Y}}{\beta + \bar{Y}} - 1 = 0 \rightarrow \bar{Y} = \beta \tag{6.47}$$

This means that centering the weighting function at the design point will yield the smallest variance for the estimated failure probability. For a value of $\beta = 3.0$ the variance is reduced by a factor of 164 as compared to plain Monte Carlo simulation which means that the computational effort to obtain a certain level of confidence is reduced substantially.

Importance sampling at the design point

Based on the previous FORM analysis it may be attempted to obtain a general importance sampling concept. This can be accomplished in two steps:

1. Determine the design point $\mathbf{x}^*$ as shown in the context of the FORM-procedure.
2. Choose a weighting function (sampling density) $h_{\mathbf{Y}}(\mathbf{x})$ with the statistical moments $E[\mathbf{Y}] = \mathbf{x}^*$ and $\mathbf{C}_{\mathbf{YY}} = \mathbf{C}_{\mathbf{XX}}$ in the following form (multi-dimensional Gaussian distribution, cf. Fig. 6.5)

$$h_{\mathbf{Y}}(\mathbf{x}) = \frac{1}{(2\pi)^{\frac{n}{2}}\sqrt{\det \mathbf{C}_{\mathbf{XX}}}} \exp\left[-\frac{1}{2}(\mathbf{x} - \mathbf{x}^*)^T \mathbf{C}_{\mathbf{XX}}^{-1}(\mathbf{x} - \mathbf{x}^*)\right] \tag{6.48}$$

3. Perform random sampling and statistical estimation according to Eq. 6.41.

The efficiency of this concept depends on the geometrical shape of the limit state function. In particular, limit state functions with high curvatures or almost circular shapes cannot be covered very well.

It is also interesting to note that the concept of importance sampling can very well be extended for application in the context of dynamic problems (first passage failure, Macke and Bucher 2003).

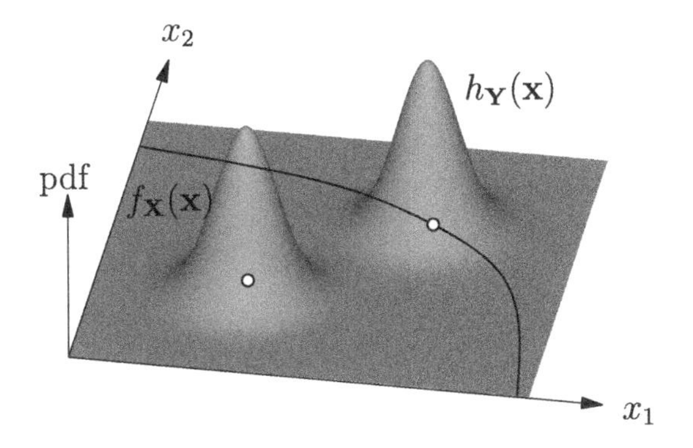

Figure 6.5 Original and importance sampling probability density functions.

Example 6.4 (Linear limit state function)
Consider a reliability problem described by two independent standardized Gaussian random variables X_1 and X_2. Let failure be defined by the limit state function

$$g(X_1, X_2) = 3 - X_1 - X_2 \tag{6.49}$$

The design point for this problem is located at $\mathbf{x}^* = [1.5, 1.5]^T$. An octave-script carrying out both crude Monte Carlo and importance sampling at the design point is given in Listing 6.2.

```
1   function limit=limit(x1,x2)
2     limit = 3-x1-x2;
3     endfunction
4   function weight=weight(x1,x2,shift)
5     weight=exp(-x1*shift(1)+shift(1)^2/2-x2*shift(2)+
        shift(2)^2/2);
6     endfunction
7   NSIM=200
8   despo=[1.5,1.5]
9   x1=randn(NSIM,1);
10  x2=randn(NSIM,1);
11  f=limit(x1,x2);
12  fail=sum(f<0)/NSIM
13  y1=x1+despo(1);
14  y2=x2+despo(2);
15  g=limit(y1,y2);
16  ifail=(g<0).*weight(y1,y2,despo);
17  ifail=sum(ifail)/NSIM
18  fid1=fopen("plainsafe.txt","w")
19  fid2=fopen("plainfail.txt","w")
20  fid3=fopen("impsafe.txt","w")
21  fid4=fopen("impfail.txt","w")
22  for i=1:NSIM
23    if (f(i)>0) fprintf(fid1,"%g %g\n", x1(i),x2(i));
24    else  fprintf(fid2,"%g %g\n", x1(i),x2(i));
25    endif
26    if (g(i)>0) fprintf(fid3,"%g %g\n", y1(i),y2(i));
27    else  fprintf(fid4,"%g %g\n", y1(i),y2(i));
28    endif
29    endfor
30  fclose(fid1);
31  fclose(fid2);
32  fclose(fid3);
33  fclose(fid4);
```

Listing 6.2 Importance sampling at the design point.

Running this code yields the results

```
1 fail = 0.015000
2 ifail = 0.017493
```

The corresponding 200 sample points are shown in Fig. 6.6 for crude Monte Carlo as well as for importance sampling. The exact result for this problem is $p_f = \Phi(-1.5\sqrt{2}) = 0.01695$

Exercise 6.1 (Reliability problem with one Gaussian and one non-Gaussian variable) Consider a simple reliability problem with $g(X_1, X_2) = X_1 - X_2$ and let X_1 be Gaussian with $\bar{X}_1 = 1$, $\sigma_{X_1} = 0.4$ and X_2 be exponentially distributed with $\bar{X}_2 = \sigma_{X_2} = 0.2$. X_1 and X_2 are assumed to be stochastically independent. Calculate p_f based on importance sampling using the design point.

Result: The resulting failure probability is $p_f = 0.0359$ as given in Example 6.2.

Adaptive sampling

As mentioned earlier, the "optimal" sampling density should satisfy the requirement

$$h_\mathbf{Y}(\mathbf{x}) = f_\mathbf{X}(\mathbf{x}|\mathbf{x} \in D_f) \tag{6.50}$$

Here the failure domain D_f is the set in which the limit state function is negative

$$D_f = \{\mathbf{x}|g(\mathbf{x}) \leq 0\} \tag{6.51}$$

This ideal condition cannot be met strictly. Yet it is possible to meet it in a second moment sense, i.e. $h_\mathbf{Y}(\mathbf{x})$ can be chosen such that (Bucher 1988a)

$$\mathbf{E}[\mathbf{Y}] = \mathbf{E}[\mathbf{X}|\mathbf{X} \in D_f] \tag{6.52}$$

$$\mathbf{E}[\mathbf{Y}\mathbf{Y}^T] = \mathbf{E}[\mathbf{X}\mathbf{X}^T|\mathbf{X} \in D_f] \tag{6.53}$$

In terms of these statistical moments, a multi-dimensional Gaussian distribution is uniquely determined.

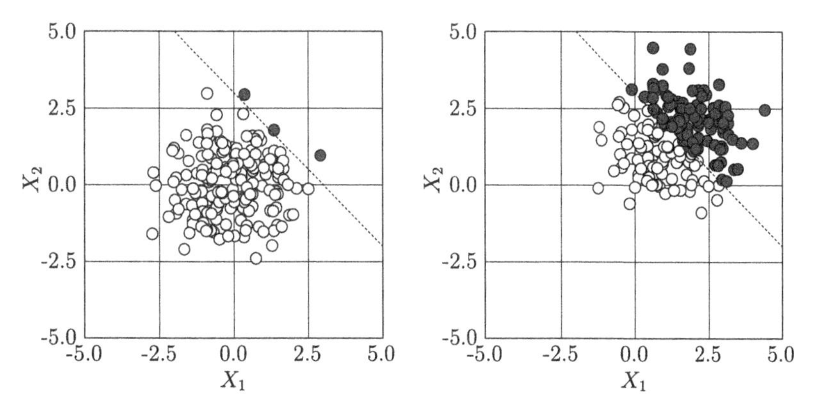

Figure 6.6 Crude Monte Carlo sampling (left) and importance sampling (right).

Example 6.5 (1-Dimensional problem)
Consider the 1-dimensional problem discussed above in which X is a Gaussian random variable with zero mean and unit variance, and the limit state functions is $g(x) = \beta - x$. It can easily be seen that

$$p_f = \Phi(-\beta) \tag{6.54}$$

The conditional mean of X in the failure domain is

$$\begin{aligned}\mathbf{E}[Y] = \mathbf{E}[X|X \in D_f] &= \frac{1}{p_f}\int_\beta^\infty \frac{x}{\sqrt{2\pi}}\exp\left(-\frac{x^2}{2}\right)\mathrm{d}x \\ &= \frac{1}{\sqrt{2\pi}}\frac{1}{\Phi(-\beta)}\exp\left(-\frac{\beta^2}{2}\right)\end{aligned} \tag{6.55}$$

For large values of β this approaches the value of β. The conditional second moment in the failure domain becomes

$$\begin{aligned}\mathbf{E}[Y^2] &= \mathbf{E}[X^2|X \in D_f] \\ &= \frac{1}{p_f}\int_\beta^\infty \frac{x^2}{\sqrt{2\pi}}\exp\left(-\frac{x^2}{2}\right)\mathrm{d}x = 1 + \beta\mathbf{E}[Y]\end{aligned} \tag{6.56}$$

and the corresponding variance is

$$\sigma_Y^2 = 1 + \beta\mathbf{E}[Y] - \mathbf{E}[Y]^2 \tag{6.57}$$

We now perform a series expansion for large values of β. According to Abramowitz and Stegun 1970 we have

$$\Phi(-\beta) = \frac{1}{\sqrt{2\pi}}\frac{1}{\beta}\exp\left(-\frac{\beta^2}{2}\right)\left[1 - \frac{1}{\beta^2}\ldots\right] \tag{6.58}$$

so that

$$\bar{Y} \approx \exp\left(-\frac{\beta^2}{2}\right)\beta\frac{\beta^2}{\beta^2 - 1}\exp\left(\frac{\beta^2}{2}\right) = \frac{\beta^3}{\beta^2 - 1} \tag{6.59}$$

which tends to $\bar{Y} = \beta$ as $\beta \to \infty$. Using the same asymptotic expansion for $\Phi(-\beta)$ we get

$$\sigma_Y^2 \approx 1 + \beta\frac{\beta^3}{\beta^2 - 1} - \frac{\beta^6}{(\beta^2 - 1)^2} = 1 - \frac{\beta^4}{(\beta^2 - 1)^2} \tag{6.60}$$

which tends to 0 as $\beta \to \infty$. Again, this confirms that the major contribution to the failure probability comes from the immediate vicinity of the design point. Table 6.1 shows numerical values of $\bar{Y}$ and σ_Y for different values of β.

While, of course, this approach cannot be applied from the beginning without prior knowledge, it is fairly easy to estimate $\mathbf{E}[X|X \in D_f]$ and $\mathbf{E}[X^2|X \in D_f]$ from a pilot simulation (based on e.g. the knowledge of the design point or by using an increased

Table 6.1 Conditional mean and standard deviation in the failure domain.

β	$\bar{Y}$	σ_Y
0	0.798	0.603
1	1.525	0.447
2	2.368	0.358
3	3.283	0.266
4	4.222	0.254

sampling standard deviation to increase the number of samples in the failure domain), and then to adapt the sampling density according to these results.

Although this adaptive approach has been found to be quite efficient in numerous applications, a closer inspection of the variance of the estimator for p_f reveals a substantial theoretical problem. In order to demonstrate this problem by means of a one-dimensional example, consider the case $\beta = 3$ in Table 6.8. For this case, the Gaussian sampling density constructed from $\bar{Y}$ and σ_Y has the form

$$h_Y(x) = \frac{1}{\sqrt{2\pi} \cdot 0.266} \exp\left[-\frac{(x-3.282)^2}{2 \cdot 0.266^2}\right] \tag{6.61}$$

The standard deviation of the estimated failure probability can then be computed from

$$\begin{aligned} m \cdot \sigma^2_{\bar{p}_f} &= \int_3^\infty \left[\frac{f_X(x)}{h_Y(x)}\right]^2 h_Y(x)\mathrm{d}x = 0.266^2 \int_3^\infty \exp\left[-x^2 + \frac{(x-3.282)^2}{2 \cdot 0.266^2}\right]\mathrm{d}x \\ &= 0.266^2 \int_3^\infty \exp\left(6.0665x^2 - 17.4151x + 76.1174\right)\mathrm{d}x = \infty \end{aligned} \tag{6.62}$$

Hence the variance is clearly unbounded. Nevertheless, the method works in practice. How can this be explained? The reason is due to numerical limitations. Random number generators for Gaussian variables cannot produce infinite values. Hence the numbers will be bounded in some interval about the mean, say $Y \in [\bar{Y} - \alpha\sigma_Y, \bar{Y} + \alpha\sigma_Y]$. If we assume a symmetrically truncated Gaussian distribution over this interval for $h_Y(y)$, then the expected value of the estimator will be

$$\begin{aligned} E[\bar{p}_f] &= \int_{\max(3,\bar{Y}-\alpha\sigma_Y)}^{\bar{Y}+\alpha\sigma_Y} \frac{f_X(x)}{h_Y(x)} h_Y(x)\mathrm{d}x = \int_{\max(3,\bar{Y}-\alpha\sigma_Y)}^{\bar{Y}+\alpha\sigma_Y} f_X(x)\mathrm{d}x \\ &= \Phi(\max(3, \bar{Y} - \alpha\sigma_Y)) - \Phi(\bar{Y} + \alpha\sigma_Y) \end{aligned} \tag{6.63}$$

Clearly, this will not be equal to the true value of $p_f = \Phi(-3)$, i.e. the estimator is biased. The dependence of the bias on α is shown in Fig. 6.7. It can be seen that for values of $\alpha > 3$ there is no substantial bias.

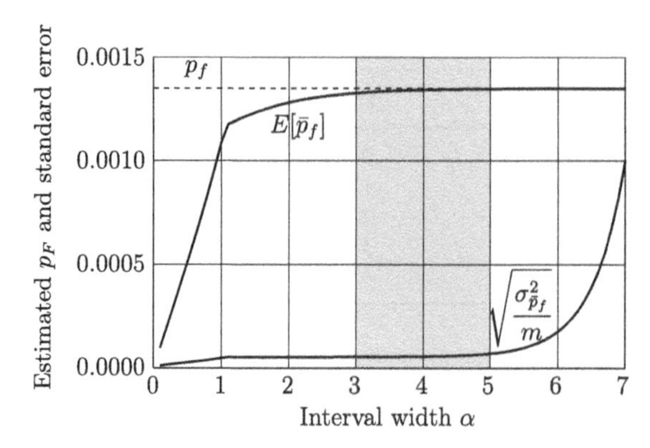

Figure 6.7 Expected value and standard error (for $m = 1000$) of estimated failure probability vs. truncation interval width α.

The variance of the estimator can be computed from

$$m \cdot \sigma^2_{\bar{p}_f} = \int_{\max(3, \bar{Y} - \alpha\sigma_Y)}^{\bar{Y} + \alpha\sigma_Y} \left[\frac{f_X(x)}{h_Y(x)}\right]^2 h_Y(x)\mathrm{d}x = \int_{\max(3, \bar{Y} - \alpha\sigma_Y)}^{\bar{Y} + \alpha\sigma_Y} \frac{f_X(x)^2}{h_Y(x)}\mathrm{d}x \tag{6.64}$$

Results are given in Fig. 6.7. In this figure, for better comparison the standard error has been computed according to

$$S_E = \sqrt{\frac{\sigma^2_{\bar{p}_f}}{m}} \tag{6.65}$$

in which $m = 1000$ has been chosen for the number of samples. It can be observed that for values $\alpha < 5$ the standard error is very small. So considering the expected value (related to bias) and the standard error (related to consistency), a good choice for α would be in the range $3 < \alpha < 5$. Actually, using $\alpha = 4$, the coefficient of variation for the estimation of the failure probability with 1000 samples is 0.042. Compared to plain Monte Carlo with a coefficient of variation of 0.861 this is a very substantial improvement. For Monte Carlo, the number of samples required to achieve the same coefficient of variation as with adaptive sampling is $m = 420.000$.

Example 6.6 (Linear limit state function)
Consider a reliability problem described by two independent standardized Gaussian random variables X_1 and X_2. Let failure be defined by the limit state function

$$g(X_1, X_2) = 3 - X_1 - X_2 \tag{6.66}$$

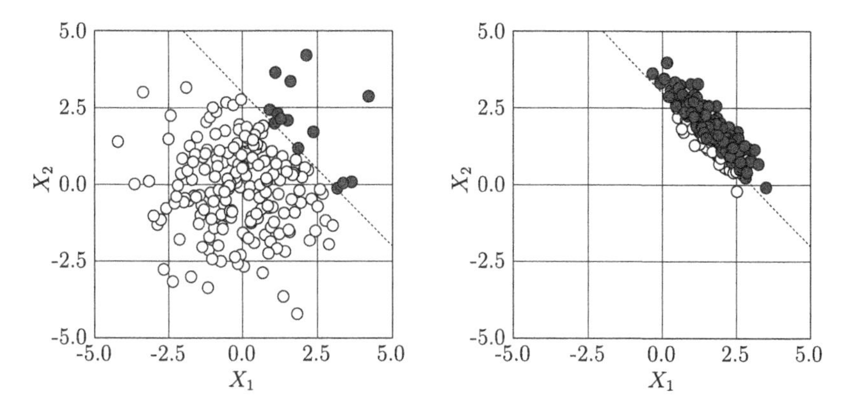

Figure 6.8 Crude Monte Carlo sampling with increased standard deviation (left) and sampling with adapted sampling density (right).

In order to achieve sufficient confidence in estimating the mean and covariance conditional on the failure domain as required for adapting the sampling density, it is useful to initially increase the standard deviation of the sampling density as compared to the original density. In this example, an initial factor of 1.5 was chosen. The corresponding 200 sample points are shown in the left part of Fig. 6.8. Estimation of the conditional mean and covariance matrix in the failure domain yields

$$\mathbf{E}[\mathbf{Y}|\mathbf{Y} \in D_f] = \begin{bmatrix} 1.588 \\ 1.792 \end{bmatrix}; \quad \mathbf{E}[\mathbf{Y}\mathbf{Y}^T|\mathbf{Y} \in D_f] = \begin{bmatrix} 0.466 & -0.434 \\ -0.434 & 0.568 \end{bmatrix} \tag{6.67}$$

in which the covariance matrix implies a coefficient of correlation for the sampling density of $\rho = -0.84$. After carrying out the adaptation, the next set of samples is significantly closer to the limit state function. These samples are shown in the right part of Fig. 6.8. The resulting failure probability for this run is $p_F = 0.0168$ which is very close to the exact result of 0.01695. Further adaptations do not change the result significantly.

6.2.3 *Directional sampling*

The basic idea is to simulate directions instead of points, and to solve analytically for the probability of failure conditional on a certain direction. The formulation is based on a representation of the limit state function in standard normal space (denoted by the random vector $\mathbf{U}$). Each point $\mathbf{u}$ in this space is written in the form of

$$\mathbf{u} = r\mathbf{a} \tag{6.68}$$

in which r is the distance from the origin and $\mathbf{a}$ is a unit vector indicating the direction. This implies transformation to n-dimensional spherical coordinates.

Density function in spherical coordinates

From $f_{\mathbf{U}}(\mathbf{u}) = \frac{1}{(2\pi)^{\frac{n}{2}}} \exp(-\frac{1}{2}\mathbf{u}^T\mathbf{u})$ we want to find the joint density $f_{R,\mathbf{A}}(r, \mathbf{a})$ of distance R and direction vector $\mathbf{A}$.

1) $f_{\mathbf{U}}(\mathbf{u})$ is rotationally symmetric, i.e. independent of $\mathbf{a}$. This follows from

$$\mathbf{u}^T\mathbf{u} = (r\mathbf{a})^T(r\mathbf{a}) = r^2\mathbf{a}^T\mathbf{a} = r^2 \tag{6.69}$$

and implies

$$f_{R|\mathbf{A}}(r|\mathbf{a}) = f_R(r) \tag{6.70}$$

This in turn yields independence of R and $\mathbf{A}$

$$f_{R,\mathbf{A}}(r, \mathbf{a}) = f_{R|\mathbf{A}}(r|\mathbf{a}) f_{\mathbf{A}}(\mathbf{a}) = f_R(r) f_{\mathbf{A}}(\mathbf{a}) \tag{6.71}$$

2) Due to rotational symmetry, $f_{\mathbf{A}}(\mathbf{a})$ must have identical values for any $\mathbf{a}$. Hence this density is constant. Its value is the inverse of the surface area of the n-dimensional unit sphere

$$f_{\mathbf{A}}(\mathbf{a}) = \frac{1}{S_n} = \frac{\Gamma(\frac{n}{2})}{2\pi^{\frac{n}{2}}} \tag{6.72}$$

For $n = 2$ we have $f_A(a) = \frac{\Gamma(1)}{2\pi} = \frac{1}{2\pi}$ and for $n = 3$ we get $f_A(a) = \frac{\Gamma(1.5)}{2\pi^{\frac{3}{2}}} = \frac{1}{4\pi}$

3) The density of $r = \sqrt{\mathbf{u}^T\mathbf{u}}$ is determined from integrating the joint density of the components of $\mathbf{u}$ over a sphere with radius r leading to

$$f_R(r) = S_n r^{n-1} \frac{1}{\pi^{\frac{n}{2}}} \exp\left(-\frac{r^2}{2}\right) \tag{6.73}$$

For the case $n = 2$ we obtain the density function

$$f_R(r) = r \exp\left(-\frac{r^2}{2}\right) \tag{6.74}$$

which describes a Rayleigh distribution.

Probability of failure

The failure probability $P(\mathcal{F}|\mathbf{a})$ conditional on a realization of the direction $\mathbf{a}$ can be determined analytically

$$\begin{aligned} P(\mathcal{F}|\mathbf{a}) &= \int_{R^*(\mathbf{a})}^{\infty} f_{R|\mathbf{A}}(r|\mathbf{a}) dr \\ &= S_n r^{n-1} \frac{1}{\pi^{\frac{n}{2}}} \exp\left(-\frac{r^2}{2}\right) dr = 1 - \chi_n^2[R^*(\mathbf{a})^2] \end{aligned} \tag{6.75}$$

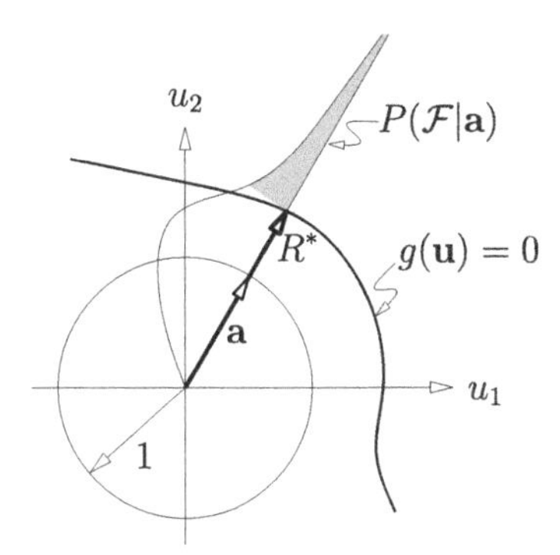

Figure 6.9 Directional sampling.

This is the cumulative Chi-Square-Distribution with n degrees of freedom.

Simulation procedure

The directional sampling procedure can be summarized as follows:

1. Generate a sample $\mathbf{u}_k$ according to an n-dimensional standard normal distribution.
2. Calculate the direction vector $\mathbf{a}_k = \frac{\mathbf{u}_k}{||\mathbf{u}_k||}$.
3. Calculate critical distance $R^*(\mathbf{a}_k)$ by solving $g[R^*(\mathbf{a}_k)\mathbf{a}] = 0$. This step may involve substantial computational effort since here the limit state function must be evaluated several time, e.g. within a bisection procedure.
4. Determine conditional failure probability $P(\mathcal{F}|\mathbf{a}) = 1 - \chi_n^2[R^*(\mathbf{a}_k)^2]$.
5. Repeat above steps with $k \to k+1$.

The method works optimally if the limit state function is circular. In this special case, one single sample yields the exact result.

Example 6.7 (Two-dimensional problem)
Consider again a reliability problem described by two independent standardized Gaussian random variables X_1 and X_2. Let failure be defined by the limit state function

$$g(X_1, X_2) = 3 - X_1 - X_2 \qquad (6.76)$$

For this limit state function, the critical distance R^* can be easily computed. Given a sample unit vector $\mathbf{a}$ with components a_1 and a_2 we immediately get

$$R^* = \frac{3}{a_1 + a_2} \qquad (6.77)$$

A directional simulation run with 200 samples is shown in Fig. 6.10. Only those points which resulted in a critical distance R^* less than 5 are shown in the figure. Also, all directions which result in a negative value of R^* have been omitted. This leaves a total of 63 directions. From this run, we obtain an estimated probability of failure of $p_f = 0.0153$. This is reasonably close to the exact result of $p_f = 0.01695$.

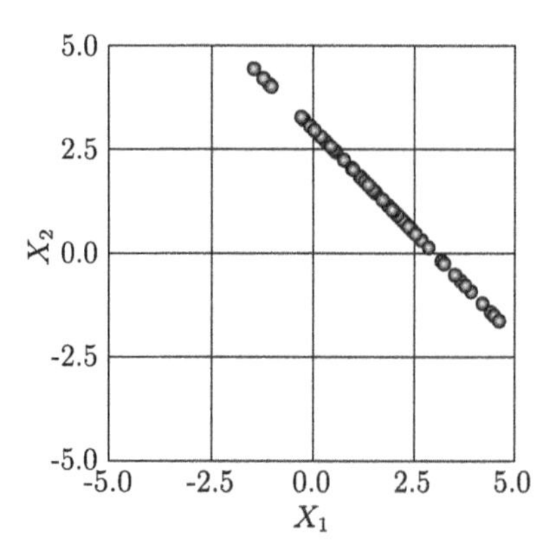

Figure 6.10 Directional sampling, points on the limit state.

```
 1  NSIM=200
 2  pf=0;
 3  fid=fopen('dirsamp.txt', 'w');
 4  for i=1:NSIM
 5    a=randn(2,1);
 6    anorm = sqrt(a(1)^2+a(2)^2);
 7    a = a/anorm;
 8    Rstar = 3/(a(1)+a(2));
 9    if (Rstar>0)
10      pf = pf + (1-chisquare_cdf(Rstar^2, 2))/NSIM;
11      if (Rstar < 5)
12        fprintf(fid, '%g %g\n', a(1)*Rstar, a(2)*Rstar);
13      end
14    end
15  end
16  fclose(fid);
17  pf
```

Listing 6.3 Directional sampling for two-dimensional linear problem.

The octave-script carrying out the directional sampling is shown in Listing 6.3.

6.2.4 *Asymptotic sampling*

General concept

The following approach has been presented in (Bucher 2008). It relies on the asymptotic behavior of the failure probability in n-dimensional i.i.d Gaussian space as the standard deviation σ of the variables and hence the failure probability P_F approaches zero (see Breitung 1984). Consider a (possibly highly nonlinear) limit state function $g(\mathbf{X})$ in which $g < 0$ denotes failure. Let σ be the standard deviation of the i.i.d. Gaussian

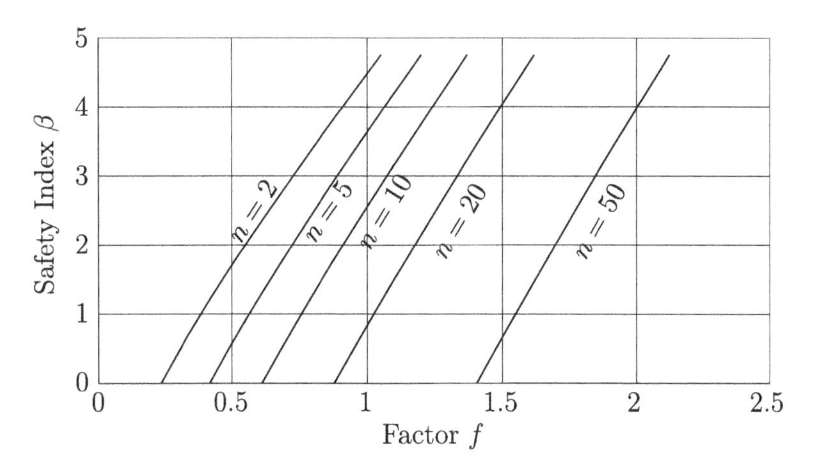

Figure 6.11 Relation between safety index and standard deviation of basic variables for hyper-circular limit state.

variables $X_k, k = 1 \dots n$. We are going to determine the functional dependence of the generalized safety index β on the standard deviation σ by using an appropriate sampling technique. This is aided by some analytical considerations involving limit cases.

First, we study the case of a linear limit state function. This problem can always be reduced to a single variable by an appropriate coordinate transformation. Hence the safety index $\beta(\sigma)$ is simply given by

$$\beta(\sigma) = \frac{\beta(1)}{\sigma} \tag{6.78}$$

in which $\beta(1)$ is the safety index evaluated for $\sigma = 1$. Introducing the scale variable $f = \frac{1}{\sigma}$ we obtain the linear relation

$$\beta(f) = f \cdot \beta(1) \tag{6.79}$$

This means that in order to obtain a good estimate for $\beta(1)$, we can compute the safety index for a larger value of σ (corresponding to a smaller value of the scale f) using Monte Carlo simulation and then simply extrapolate by multiplying the obtained result with f (i.e. divide by σ).

As the second analytical case, consider the other extreme case of a (hyper)circular limit state function in n-dimensional Gaussian space in which failure is given by $g(\mathbf{X}) = R^2 - \mathbf{X}^T\mathbf{X} \leq 0$. In this case, the probability of failure depending on the scale f is given in terms of the χ^2-distribution with n degrees of freedom

$$P_F = 1 - \chi^2(f^2R^2, n) \rightarrow \beta = \Phi^{-1}\left[1 - \chi^2(f^2R^2, n)\right] \tag{6.80}$$

This relationship between β and f is shown in Fig. 6.11 for $R = 5$ and for various values of n. It can be seen that the relation approaches a linear one with increasing n and increasing β. The latter is in agreement with the well-established result that the

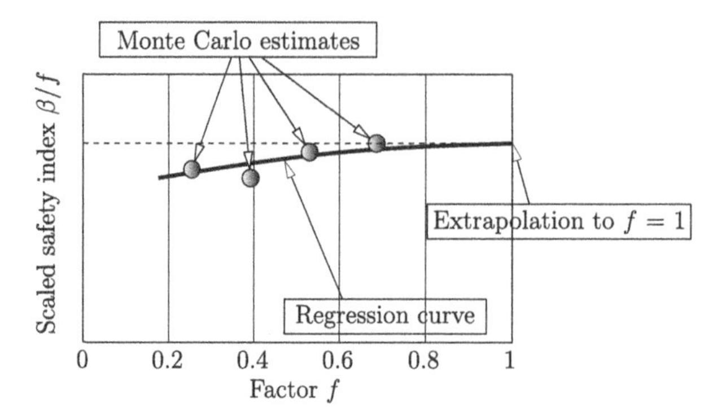

Figure 6.12 Basic concept of asymptotic sampling.

second order reliability method yields asymptotically exact results as $\beta \to \infty$ (Breitung 1984).

The concept of asymptotic sampling utilizes the asymptotic behavior of the safety index β by applying an extrapolation technique. Here the (assumed) functional dependence for β is chosen as

$$\beta = A \cdot f + \frac{B}{f} \tag{6.81}$$

This choice is motivated in order to ensure asymptotically linear behavior as $f \to \infty$ (which is equivalent to $\sigma \to 0$). The coefficients A and B are conveniently determined from a least-squares fit using Monte Carlo estimates of β for different values of f (typically for values of $f < 1$) as support points. For this fitting process, Eq. 6.81 is rewritten in terms of a scaled safety index as

$$\frac{\beta}{f} = A + \frac{B}{f^2} \tag{6.82}$$

This is illustrated qualitatively in Fig. 6.12. One major advantage of this approach is its independence of the dimensionality. The accuracy is governed only by the relation between the number of samples and the probability of failure as well as the particular geometry of the limit state surface $g(\mathbf{u}) = 0$.

In this context it is essential to use a sampling method which provides very stable results. One obvious choice is Latin Hypercube Sampling (LHS) (Imam and Conover 1982; Florian 1992). Alternatively, pseudo-random sequences with low discrepancy (Niederreiter 1992; Sobol and Asotsky 2003) can be utilized. In the following examples randomized Sobol sequences using an algorithm as discussed in Bratley and Fox 1988 and Hong and Hickernell 2003 are used.

Example 6.8 (Simple limit state function, subtraction of random variables)
As a first example, consider a simple two-dimensional limit state function $g = X_1 - X_2$ in which X_1 is lognormally distributed with a mean value $\bar{X}_1 = 2$, a standard deviation $\sigma_1 = 0.2$ and X_2 normal with mean value $\bar{X}_2 = 1$ and standard deviation $\sigma_2 = 0.15$. The

Table 6.2 Statistics of asymptotic sampling results for different number of sample points, 1000 repetitions.

m	*LHS*		*Sobol*	
	β	σ_β	β	σ_β
128	4.269	0.448	4.347	0.316
256	4.283	0.339	4.363	0.177
1024	4.344	0.233	4.365	0.088

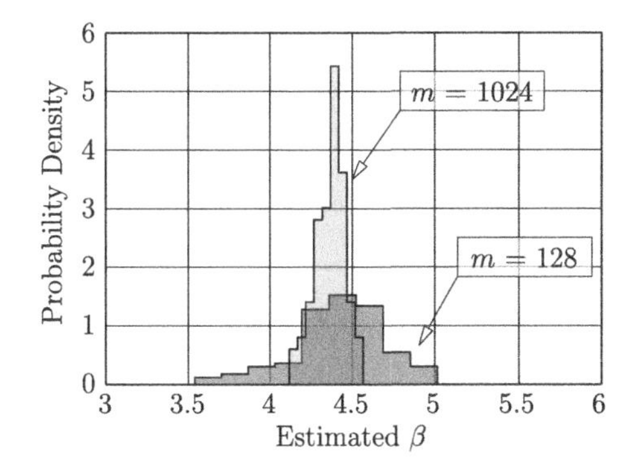

Figure 6.13 Histogram of asymptotic sampling results based on randomized Sobol sequence.

exact failure probability as obtained from Monte Carlo simulation with 10 million samples is $P_F = 5.5 \cdot 10^{-6} (\beta = 4.397)$.

The safety indices as determined from asymptotic sampling using 5 support points each of which is based on m samples are given in Table 6.2. These results were obtained using Latin Hypercube samples as support points as well as using support points generated by randomized Sobol sequences. It can be seen that the Sobol sequences yield smaller estimation errors as expressed in terms of the standard deviation of the safety indices.

A visual impression of the sampling scatter is provided by the corresponding histograms in Fig. 6.13.

The results of one run using Sobol sequences with m support points is shown in Fig. 6.14. Here it becomes quite clear that with increasing number of support points m on the one hand the dispersion of the estimated for the scaled safety index β/f is reduced and on the other hand the support points lie closer to the limit $f = 1$. The latter is due to the well-known fact that smaller failure probabilities can be estimated reliably only with a larger number of support points.

Example 6.9 (Simple limit state function, multiplication of random variables)
This example is taken from Shinozuka 1983. It involves two random variables, X_1 lognormal with mean 38 and standard deviation 3.8, and X_2 normal with mean 54 and standard deviation 2.7. The limit state function is $g(X_1, X_2) = X_1 \cdot X_2 - 1140$. The

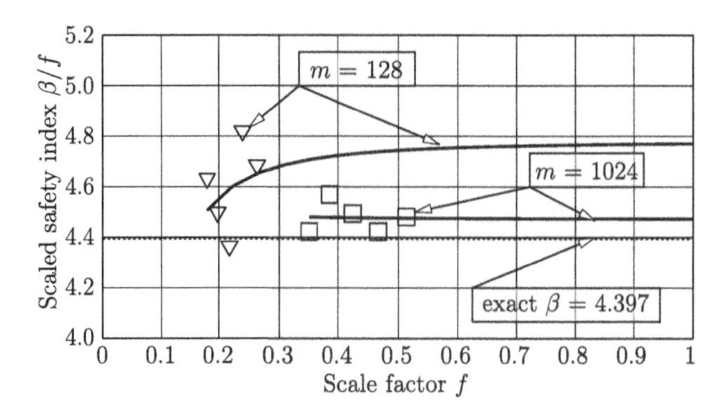

Figure 6.14 Asymptotic sampling results for example 1 based on randomized Sobol sequence.

Table 6.3 Statistics of asymptotic sampling results for different number of sample points, example 2.

m	*Sobol*	
	β	σ_β
128	4.997	0.311
256	5.044	0.210
512	5.123	0.177
1024	5.094	0.112
2048	5.118	0.072

results from asymptotic sampling using randomized Sobol sequences with different numbers m of support points are given in Table 6.3.

The exact result as reported in Shinozuka 1983 is $\beta = 5.151$, which is a result obtained from the First-Order Reliability Method (FORM).

Example 6.10 (High-dimensional linear problem)
This example serves as a test case to demonstrate the independence of the dimensionality. The limit state function is

$$g(\mathbf{X}) = 5\sqrt{n} - \sum_{k=1}^{n} X_k \tag{6.83}$$

in which n is the number of random variables. All random variables are i.i.d. standard Gaussian. The problem has a safety index of $\beta = 5$ or $P_F = 3 \cdot 10^{-7}$, independent of n. Table 6.4 shows the mean values and standard deviations of the safety index as computed from asymptotic sampling (20 repetitions with 1000 Monte Carlo samples each) for different dimension n. The `octave`-script performing this simulation is shown in Listing 6.4.

Table 6.4 Statistics of estimated safety index for high-dimensional linear problem.

n	$\bar{\beta}$	σ_β
10	4.95	0.26
100	4.94	0.22
1000	4.95	0.24
10000	4.94	0.22
100000	5.00	0.23

```
1  nvar=100000
2  nsim=1000
3  nrep=20
4  pf=0
5  f = 0.2
6  beta = []
7  for k=1:nrep
8    for i=1:nsim
9      x=randn(nvar,1)/f;
10     xlimit = 5*sqrt(nvar) - sum(x);
11     if (xlimit<0)
12       pf = pf+1;
13     endif
14     endfor
15   pf = pf/nsim
16   beta = [beta;normal_inv(1-pf)/f]
17   endfor
18 bm = mean(beta)
19 bs = std(beta)
```

Listing 6.4 Asymptotic sampling for high-dimensional problem.

6.3 Application of response surface techniques to structural reliability

6.3.1 *Basic concept*

The response surface method has been a topic of extensive research in many different application areas since the influential paper by Box and Wilson in 1951 (Box and Wilson 1951). Whereas in the initial phase the general interest was on experimental designs for polynomial models (see, e.g., Box and Wilson 1951; Box and Draper 1959), in the following years non-linear models, optimal design plans, robust designs and multi-response experiments—to name just a few—came into focus. A fairly complete review on existing techniques and research directions of the response surface methodology can be found in Hill and Hunter 1966; Mead and Pike 1975; Myers, Khuri, and W. H. Carter 1989; Myers 1999. However, traditionally the application area of the

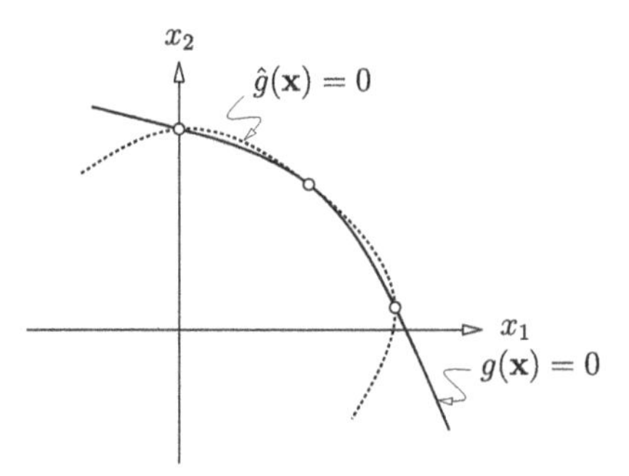

Figure 6.15 Response surface approximation for reliability analysis.

response surface method is not structural engineering, but, e.g., chemical or industrial engineering. Consequently, the above mentioned special requirement for structural reliability analysis—i.e., the high degree of accuracy required in a very narrow region—is usually not reflected upon in the standard literature on the response surface method (Box and Draper 1987; Khuri and Cornell 1996; Myers and Montgomery 2002).

One of the earliest suggestions to utilize the response surface method for structural reliability assessment was made in Rackwitz 1982. Therein, Lagrangian interpolation surfaces and second-order polynomials are rated as useful response surfaces. Moreover, the importance of reducing the number of basic variables and error checking is emphasized. Support points for estimating the parameters of the response surface are determined by spherical design. In Wong 1985 first-order polynomials with interaction terms are utilized as response surfaces to analyze the reliability of soil slopes. The design plan for the support points is saturated—either by full or by fractional factorial design. Another analysis with a saturated design scheme is given in Bucher and Bourgund 1990, where quadratic polynomials without interaction terms are utilized to solve problems from structural engineering. Polynomials of different order in combination with regression analysis are proposed in Faravelli 1989, whereby fractional factorial designs are utilized to obtain a sufficient number of support points. The validation of the chosen response surface model is done by means of analysis of variance. In Ouypornprasert, Bucher, and Schuëller 1989 it has been pointed out that, for reliability analysis, it is most important to obtain support points for the response surface very close to or exactly at the limit state $g(\mathbf{x}) = 0$ (cf. Fig. 6.15). This finding has been further extended in Kim and Na 1997; Zheng and Das 2000.

Apart from polynomials of different order, piecewise continuous functions such as hyperplanes or simplices can also be utilized as response surface models. For the class of reliability problems defined by a convex safe domain, secantial hyperplane approximations such as presented by Guan and Melchers 1997; Roos, Bucher, and Bayer 2000 yield conservative estimates for the probability of failure. Several numerical studies indicate, however, that in these cases the interpolation method converges slowly from above to the exact result with increasing number of support points. The effort required for this approach is thereby comparable to Monte Carlo simulation based on directional sampling (Bjerager 1988).

Using a representation of random variables and limit state function in the standard Gaussian space, it can easily bee seen that there is a significant similarity

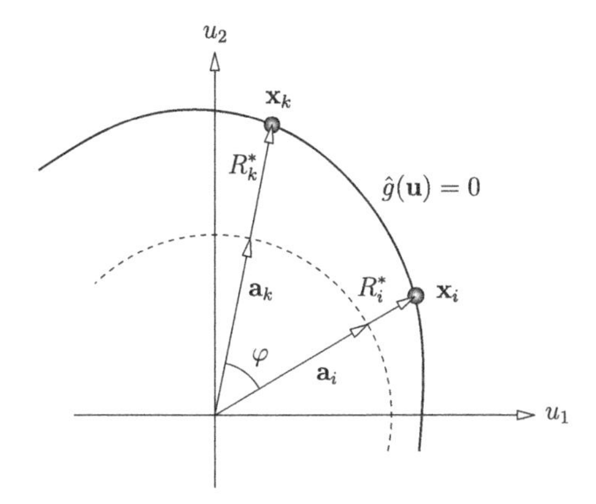

Figure 6.16 Interpolation of critical distance from origin to limit state.

between response surface method using support points at the limit state and directional sampling. The difference lies in the choice of the search directions and the way of interpolating the distance R^* from the limit state to the origin (cf. Fig. 6.16) For a useful interpolation strategy (particularly for reliability problems with convex limit state functions) we need a distance measure d which in polar coordinates is most suitable defined in terms of the angle between two vectors $\mathbf{a}_i$ and $\mathbf{a}_k$:

$$d(\mathbf{a}_i, \mathbf{a}_k) = 1 - \frac{\mathbf{a}_i^T \mathbf{a}_k}{||\mathbf{a}_i|| \cdot ||\mathbf{a}_k||} \tag{6.84}$$

which is actually one minus the cosine of the angle φ between the two vectors (cf. Fig. 6.16).

For a given direction vector $\mathbf{A}$, the distance R^* to the surface $g(\mathbf{u}) = 0$ is approximated using

$$R = \sum_{k=1}^{m} b_k \phi_k [d(\mathbf{A}, \mathbf{a}_k)] \tag{6.85}$$

in which $\phi_k(.)$ are basis functions and b_k suitable defined coefficients.

For the special case of Shepard interpolation (i.e. constant basis functions) as outlined in section 3.3.4, R^* is approximated by

$$R^* = \frac{\sum_{k=1}^{m} w_k \cdot R_k^*}{\sum_{k=1}^{m} w_k} \tag{6.86}$$

in which the weights w_k are computed from

$$w_k = \left[d(\mathbf{A}, \mathbf{a}_k) + \epsilon \right]^{-p} \tag{6.87}$$

Here, ϵ is a small positive number regularizing this expression and p is a suitably chosen positive number (e.g. $p = 2$). For reliability problems involving non-closed limit state

Table 6.5 Support points for Shepard interpolation.

i	a_{i1}	a_{i2}	R_i^*	$1/R_i^*$
1	1.000	0.000	3.000	0.333
2	0.000	1.000	3.000	0.333
3	−1.000	0.000	∞	0.000
4	0.000	−1.000	∞	0.000
5	0.707	0.707	2.121	0.471

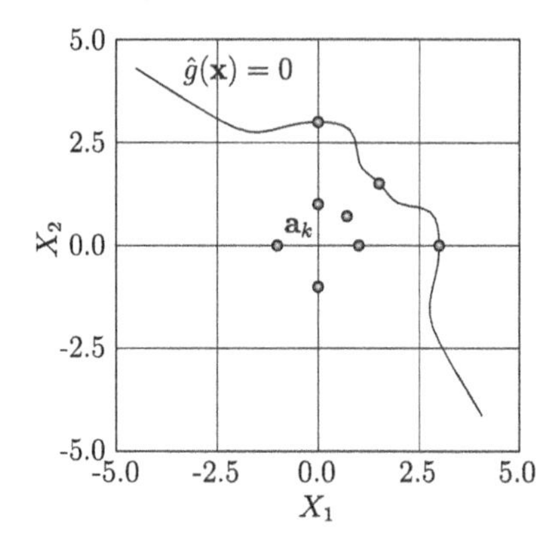

Figure 6.17 Support points and interpolating function for Shepard interpolation.

functions (i.e. problems in which R^* may become unbounded for certain directions) it may be advantageous to approximate $1/R^*$ rather than R^*:

$$\frac{1}{R^*} = \frac{\sum_{k=1}^{m} w_k \cdot 1/R_k^*}{\sum_{k=1}^{m} w_k} \tag{6.88}$$

with weights w_k as above.

Example 6.11 (Linear limit state function – Shepard interpolation)
Consider the previously discussed example of a linear limit state function in standard Gaussian space

$$g(X_1, X_2) = 3 - X_1 - X_2 \tag{6.89}$$

Using the set of direction vectors given in Table 6.5 to obtain support points we get critical distances R_i^* as shown in the same table. The octave-code generating the Shepard interpolation for this example is shown in Listing 6.5.

Carrying out a Monte-Carlo simulation run with 10000 samples based on the Shepard interpolation yields a probability of failure of $p_f = 0.0147$ which is reasonably close to the exact result pf $p_f = 0.01695$. It should be mentioned that the high quality of the approximation in this example is due to the fact that one support point for the interpolation lies exactly at the design point $\mathbf{u}^*$. This cannot be expected to happen in a more general setting.

```
1  points=[
2  1,0,1/3;
3  0,1,1/3;
4  -1,0,0;
5  0,-1,0;
6  1/sqrt(2),1/sqrt(2),1/sqrt(2*1.5^2)]
7  M=size(points)(1)
8  eps=1e-5;
9  p=2;
10 N=100
11 dphi=2*pi/(N-1)
12 fid=fopen('shep_linear_dir.txt','w');
13 for k=1:M
14   fprintf(fid, '%g %g \n', points(k,1), points(k,2));
15   end
16 fclose(fid);
17 fid=fopen('shep_linear_sup.txt','w');
18 for k=1:M
19   Rk=points(k,3);
20   if (Rk>0.15) fprintf(fid, '%g %g\n', points(k,1)/Rk,
       points (k,2)/Rk); end
21   end
22 fclose(fid);
23 %
24 fid=fopen('shep_linear.txt','w');
25 for i=1:N
26   phi=-pi+(i-1)*dphi;
27   A=[cos(phi);sin(phi)];
28   w=0;
29   R=0;
30   for k=1:M
31     ak=[points(k,1);points(k,2)];
32     Rk=points(k,3);
33     dk=1-A(1)*ak(1)-A(2)*ak(2);
34     wk = (dk+eps)^(-p);
35     R= R + Rk*wk;
36     w = w + wk;
37   end
38   R = R/w
39   if (R>0.15) fprintf(fid,'%g %g\n',A(1)/R,A(2)/R); end
40 end
41 fclose(fid);
```

Listing 6.5 Shepard interpolation for linear limit state function.

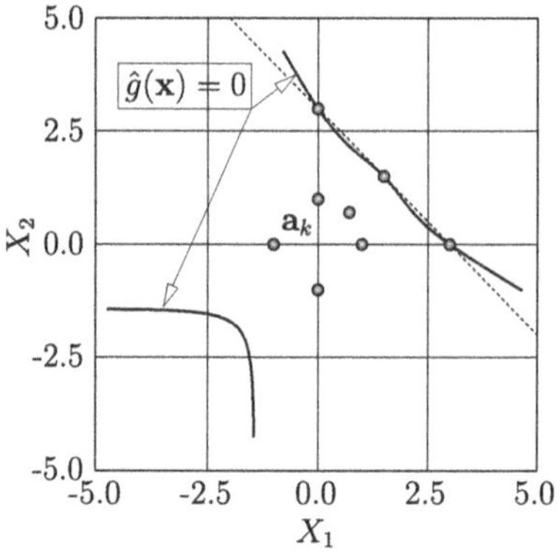

Figure 6.18 Support points and interpolating function for radial basis function interpolation.

Example 6.12 (Linear limit state function – radial basis function interpolation) Consider the previously discussed problem (cf. Example 6.11) of a linear limit state function in standard Gaussian space. Based on the same set of direction vectors and support points (cf. Table 6.5) we obtain the radial basis function interpolation as shown in Fig. 6.18. Note that the interpolation results in a second branch ("ghost" branch) of the limit state function in the third quadrant. This is an artefact of the interpolation procedure, and can be eliminated by appropriate modifications of the interpolation procedure or by including more support points. The `octave`-code generating the radial basis function (thin plate spline) interpolation (i.e. the relevant branch) for this example is shown in Listing 6.6.

Carrying out a Monte-Carlo simulation run with 10000 samples based on the radial basis interpolation yields a probability of failure of $p_f = 0.023$ which is somewhat larger than the exact result pf $p_f = 0.01695$.

6.3.2 *Structural examples*

One-bay one-storey frame

This is a simple analytical example representing the failure of a one-bay one-storey frame. The failure is assumed to be described by first-order rigid-plastic hinge theory. Due to the presence of horizontal and vertical loads (cf. Fig. 6.19) there are three relevant collapse mechanisms as shown in Fig. 6.20. The three failure modes as shown are defined by the relations

$$\begin{aligned} g_1(\mathbf{X}) &= 4\frac{M_{pl}}{L} - H; \\ g_2(\mathbf{X}) &= 4\frac{M_{pl}}{L} - V; \\ g_3(\mathbf{X}) &= 6\frac{M_{pl}}{L} - H - V; \\ \mathbf{X} &= [H, V]^T \end{aligned} \tag{6.90}$$

```
1  points=[
2  1,0,1/3;
3  0,1,1/3;
4  -1,0,0;
5  0,-1,0;
6  1/sqrt(2),1/sqrt(2),1/sqrt(2*1.5^2)]
7  M=size(points)(1)
8  eps=1e-5;
9  p=2;
10 N=100
11 dphi=2*pi/(N-1)
12 WIK=zeros(M,M);
13 for i=1:M
14   ai=[points(i,1);points(i,2)];
15   for k=1:M
16     ak=[points(k,1);points(k,2)];
17     Rk=points(k,3);
18     dik=abs(1-ai(1)*ak(1)-ai(2)*ak(2))
19     i,k,dik
20     wik=0;
21     if (dik>0) wik = dik^2*log(dik); end;
22     WIK(i,k)=wik;
23   end
24 end
25 z=points(:,3)
26 coeff=WIK\z
27 fid=fopen('radial_linear1.txt','w');
28 for i=1:N/2
29   phi=-0.25*pi+(i-1)*dphi;
30   A=[cos(phi);sin(phi)]
31   R=0;
32   for k=1:M
33     ak=[points(k,1);points(k,2)];
34     Rk=points(k,3);
35     dk=abs(1-A(1)*ak(1)-A(2)*ak(2));
36     wk = 0;
37     if (dk>0) wk = dk^2*log(dk); end;
38     R=R + wk*coeff(k);
39   end
40   R
41   if (R>0.2) fprintf(fid, '%g %g\n', A(1)/R, A(2)/R); end
42 end
43 fclose(fid);
```

Listing 6.6 Radial basis function interpolation for linear limit state function.

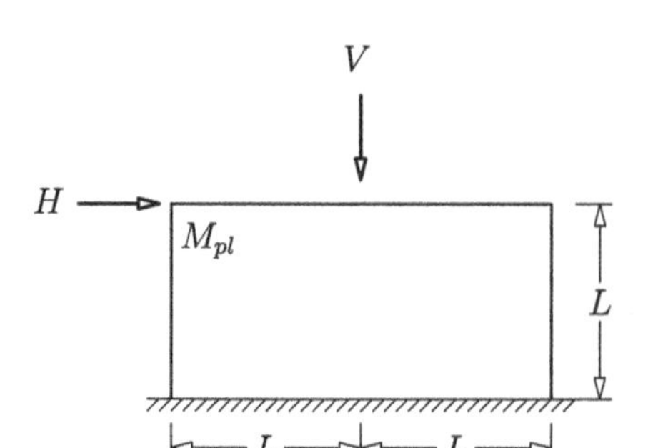

Figure 6.19 Simple frame.

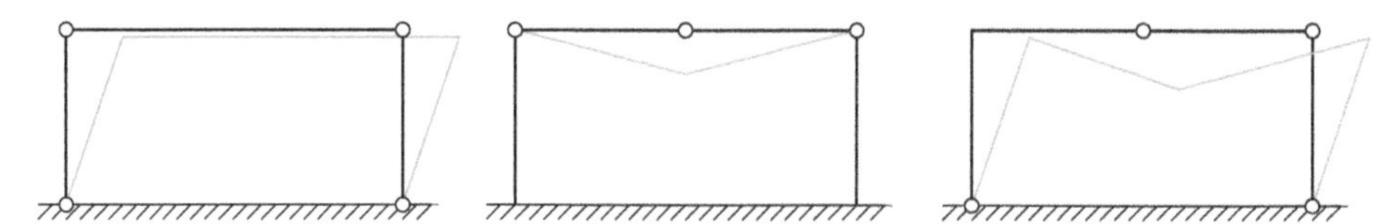

Figure 6.20 Collapse mechanisms.

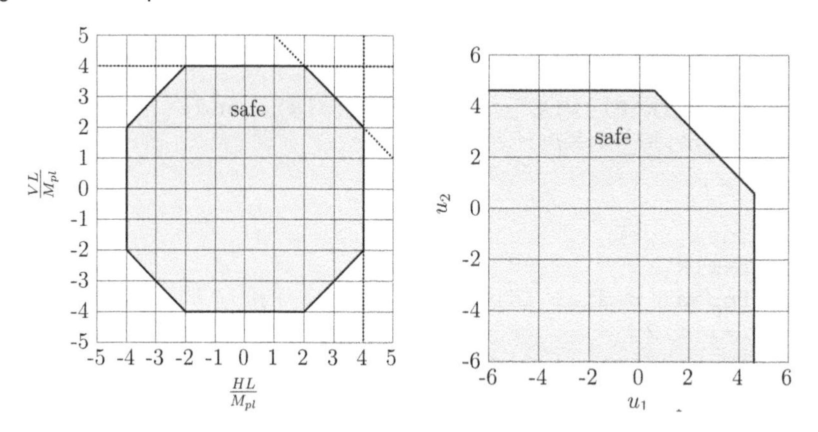

Figure 6.21 Limit state function for collapse of one-bay one-story frame in original space (left) and in standard Gaussian space (right).

The limit state function describing all possible failure modes including symmetries of the structure is shown in Fig. 6.21 (left). We assume that the plastic moment M_{pl} is deterministic, and the loads are independent Gaussian random variables with mean values of $\bar{H} = \bar{V} = 1.7\frac{M_{pl}}{L}$ and a coefficient of variation of 0.5. Hence we obtain standard Gaussian variables u_1 and u_2 by means of

$$U_1 = \frac{H - 1.7\frac{M_{pl}}{L}}{0.5\frac{M_{pl}}{L}}; \quad U_2 = \frac{V - 1.7\frac{M_{pl}}{L}}{0.5\frac{M_{pl}}{L}} \tag{6.91}$$

The representation of the limit state function in standard Gaussian space is shown in Fig. 6.21 (right). It can be observed that only a small region of the limit state function

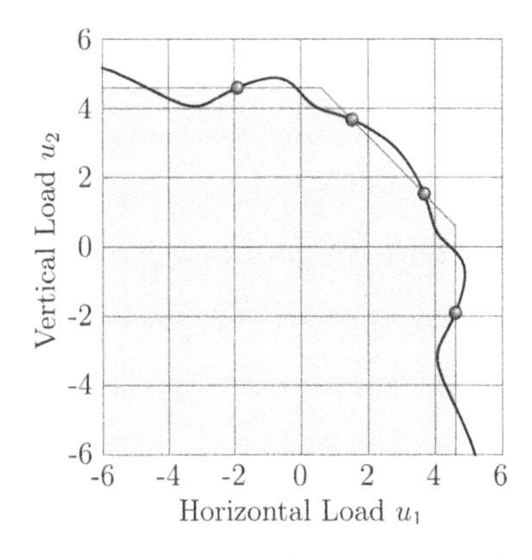

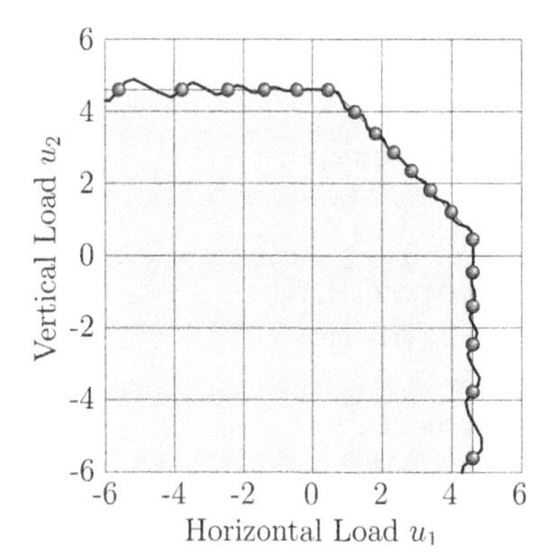

Figure 6.22 Approximate limit state using Shepard interpolation with 8 support points (left) and with 32 support points (right).

Table 6.6 Probability of failure using different approximations (simple frame).

Method	m	$p_f \cdot 10^{-5}$	β	Error in β (%)
Shepard	8	1.27	4.21	2.2
	32	1.85	4.13	0.2
RBF	8	3.50	3.98	1.3
	32	2.07	4.10	−0.5

is in the vicinity of the origin. Hence it will be most important to approximate this part well in terms of a response surface.

Using directional sampling with 10000 samples, the failure probability is determined as $p_f = 1.883 \cdot 10^{-5}$ corresponding to a value of the safety index $\beta = 4.12$. This value is used as a reference for comparison to the results based on various response surface approaches. The application of approximations to the limit state function using different approaches is compared in terms of choice of method and number of support points. The response surfaces for a different number of support points are shown in Fig. 6.22 (Shepard interpolation) and Fig. 6.23 (radial basis function interpolation). In Table 6.6, the results are shown for Shepard interpolation (Eq. 3.38) and radial basis functions (RBF, Eq. 3.49). Comparing the results for p_f it can be seen that there is not a fundamental difference in accuracy for a given number of support points, i.e. for a specific level of numerical effort. Although the geometrical representation of the limit state function differs (e.g. in the curvature), the computed probability of failure p_f — and even more so the safety index β — is not severely affected by this difference.

The `octave`-code for the directional sampling simulation based on the RBF interpolation is given in Listing 6.7.

Application to non-linear finite element structure

A simple 3-dimensional steel frame subjected to three random loadings is considered as shown in Fig. 6.24. This model has been studied previously in Bucher 1998 and Bucher and Most 2008.

```
M=32
eps=1e-5;
points=zeros(M,3);
p=2;
N=1000
WIK=zeros(M,M);
dphi=2*pi/M
SIGMA=0.5;
R=zeros(8,1);
for i=1:M
  phi=dphi/2+(i-1)*dphi;
  A=[cos(phi);sin(phi)];
  R(1)=2.3/SIGMA/A(1);
  R(2)=2.3/SIGMA/A(2);
  R(3)=-5.7/SIGMA/A(1);
  R(4)=-5.7/SIGMA/A(2);
  R(5)=2.6/SIGMA/(A(1)+A(2));
  R(6)=6/SIGMA/(A(1)-A(2));
  R(7)=6/SIGMA/(-A(1)+A(2));
  R(8)=-9.4/SIGMA/(A(1)+A(2));
  R=R+(R<0)*1500
  Rstar=min(R)
  points(i,1)=A(1);
  points(i,2)=A(2);
  points(i,3)=1/Rstar;
  end
for i=1:M
  ai=[points(i,1);points(i,2)];
  for k=1:M
    ak=[points(k,1);points(k,2)];
    Rk=points(k,3);
    dik=abs(1-ai(1)*ak(1)-ai(2)*ak(2));
    wik=0;
    if (dik>0) wik = dik^2*log(dik); end;
    WIK(i,k)=wik;
  end
end
z=points(:,3)
coeff=WIK  z
dphi=2*pi/(N-1)
pf=0
Sn=2*pi;
for i=1:N
  u=randn(2,1);
```

Listing 6.7 Directional sampling based on RBF interpolation.

```
45      un=sqrt(u(1)^2+u(2)^2);
46      A=u/un;
47      R=0;
48      for k=1:M
49         ak=[points(k,1);points(k,2)];
50         Rk=points(k,3);
51         dk=abs(1-A(1)*ak(1)-A(2)*ak(2));
52         wk = 0;
53         if (dk>0) wk = dk^2*log(dk); end;
54         R= R + wk*coeff(k);
55      end
56      Rstar = 1/R;
57      if (Rstar>0)
58         pf = pf + (1-chisquare_cdf(Rstar^2, 2))/Sn/N;
59      end
60   end
61   pf
62   beta=normal_inv(1-pf)
```

Listing 6.7 Continued

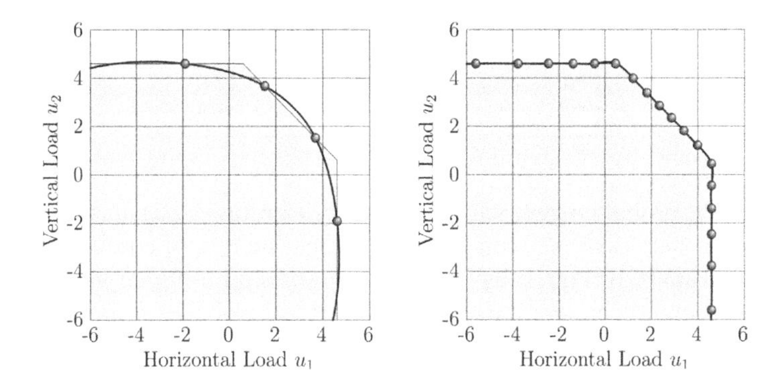

Figure 6.23 Approximate limit state using radial basis function interpolation with 8 support points (left) and with 32 support points (right).

The 3-dimensional frame is modeled by 24 physically non-linear beam elements (linear elastic-ideally plastic material law, elasticity modulus $E = 2.1 \cdot 10^{11}\text{N/m}^2$, yield stress $\sigma_Y = 2.4 \cdot 10^8\text{N/m}^2$). For numerical stabilization, a post-yielding stiffness with a magnitude equal to 10^{-5}-times the initial value is chosen. The cross section for the girder is a box (width 0.2 m, height 0.15 m, wall thickness 0.005 m) and the columns are I-sections (flange width 0.2 m, web height 0.2 m, thickness 0.005 m). The columns are fully clamped at the supports. The static loads acting on the system p_z, F_x, and F_y are assumed to be Gaussian random variables. Their respective statistical properties are given in Table 6.7. The failure condition is given by total or partial collapse of the structure. Numerically, this is checked either by tracking the smallest

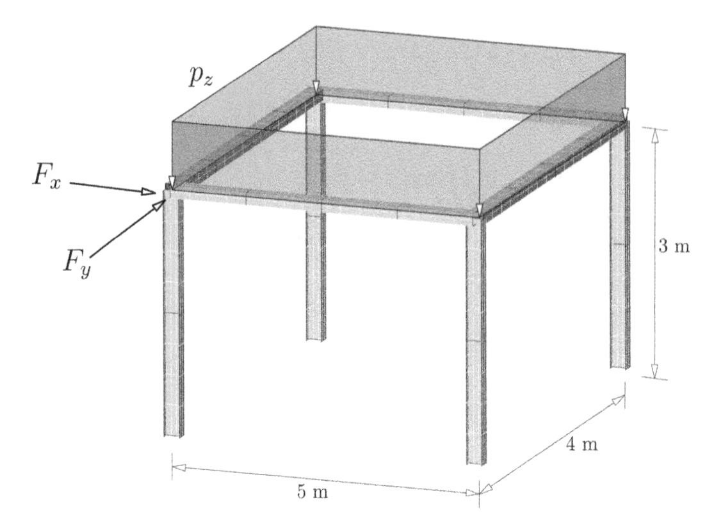

Figure 6.24 3-Dimensional steel frame structure.

Table 6.7 Random variables used in 3-dimensional frame analysis.

RV	*Mean*	*Std. Dev.*
p_z [kN/m]	12.0	1.2
F_x [kN]	30.0	3.6
F_y [kN]	40.0	4.8

eigenvalue of the global tangent stiffness matrix (it becomes zero at the collapse load) or by failure in the global Newton iteration indicating loss of equilibrium. Since this type of collapse analysis is typically based on a discontinuous function (convergence vs. non-convergence), it is imperative for the support points of the response surface to be located exactly at the limit state. A bisection procedure is utilized to determine collapse loads with high precision (to the accuracy of 1% of the respective standard deviation).

The geometry of the limit state separating the safe from the failure domain is shown in Fig. 6.25. The limit points were obtained from directional sampling using 10000 samples. It can be easily seen that there is considerable interaction between the random variables F_x and F_y at certain levels. Near the region of most importance for the probability of failure (this is where most of the points from the directional sampling are located) is essentially flat, and mainly governed by the value of p_z. The probability of failure obtained from directional sampling (10000 samples) is $p_f = 4.3 \cdot 10^{-5}$ with a standard estimation error of 3%. This corresponds to a safety index of $\beta = 3.93$.

Table 6.8 compares the results from Shepard and radial basis function interpolations using 150 and 1000 support points for each. These results were obtained using directional sampling with 10000 samples. All results are very close to the exact result obtained from directional sampling. The response surface for the Shepard model is visualized in Fig. 6.26, the response surface for the RBF model is shown in Fig. 6.27.

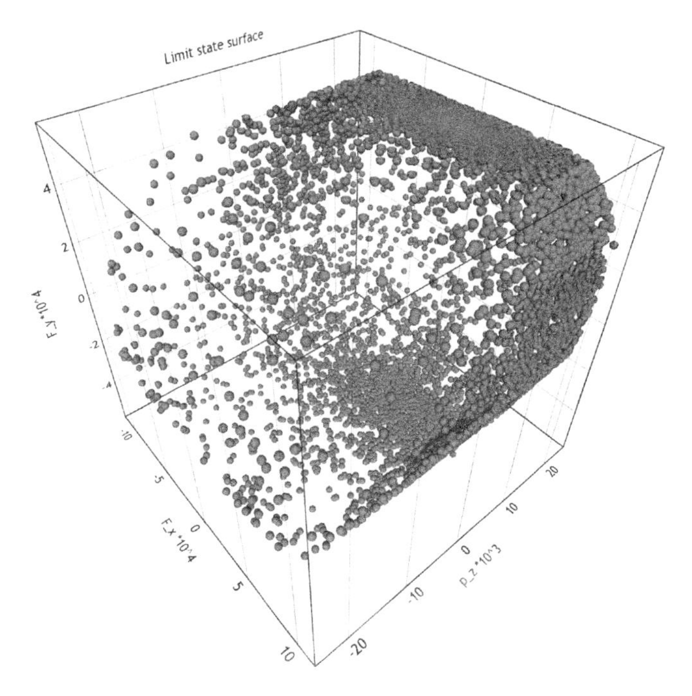

Figure 6.25 Visualization of limit state function $g(\mathbf{x})$ (directional sampling, 10.000 points).

Table 6.8 Probability of failure using different approximations (nonlinear 3D frame).

Method	m	$p_f \cdot 10^{-5}$	β	Error in β (%)
Shepard	150	2.15	4.01	2.0
	1000	3.30	3.99	1.5
RBF	150	3.84	3.95	0.5
	1000	3.36	3.99	1.5

For comparison, a saturated quadratic scheme including pairwise interactions is utilized for the layout of the experimental design scheme. The support points thus generated are interpreted as direction vectors along which all loads are incremented. Starting from the mean values and incrementing along this directions lead to a set of 9 support points on the limit state function. These support points (cf. Table 6.9) have a function value of $g(\mathbf{x}) = 0$. By adding the mean value as first support point with a function value of $g(\mathbf{x}) = 1$, a quadratic response surface can be defined. Considering lines 2–7 in Table 6.9 it has been decided to consider combination terms in which all variables are incremented up from the mean. This leads to the final three support points given in lines 8–10 of Table 6.9.

A Monte Carlo simulation based on this quadratic surface is carried out. The resulting probability of failure from plain Monte Carlo using one million samples was found

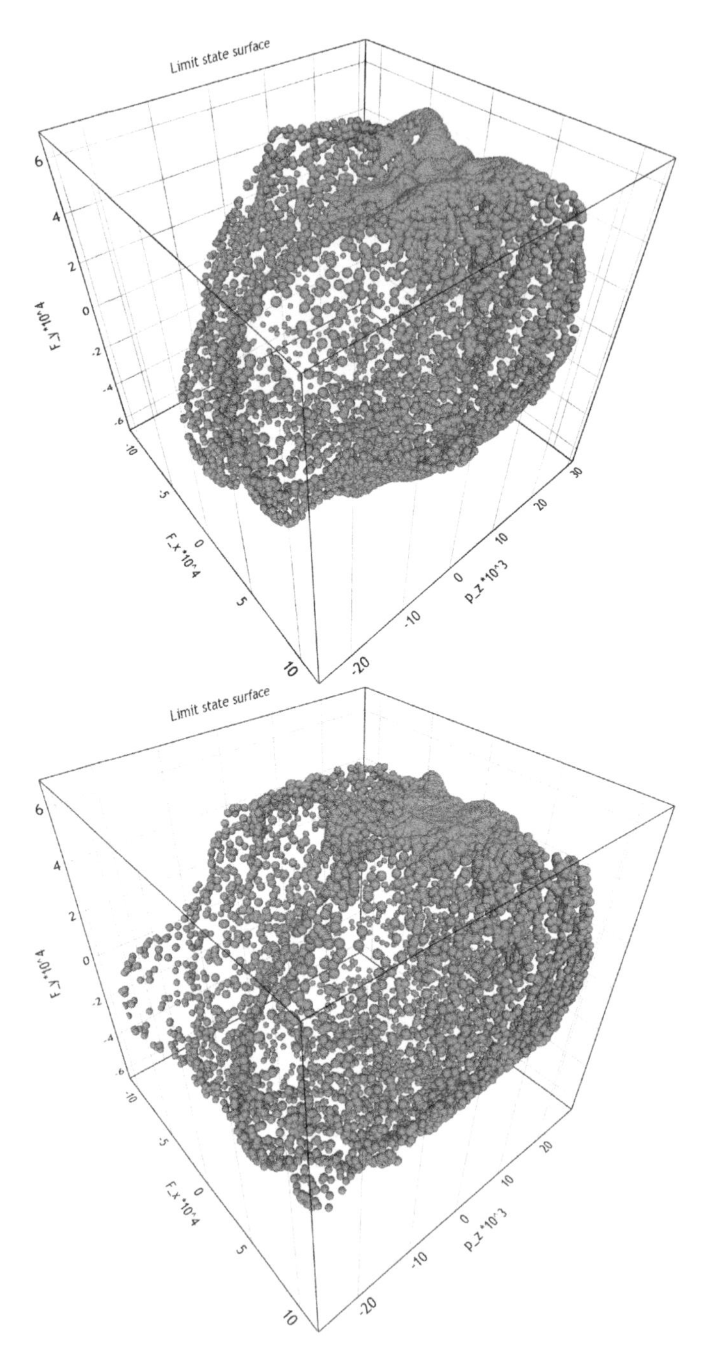

Figure 6.26 Visualization of limit state function $g(\mathbf{x})$, Shepard interpolation based on 150 points (top) and on 1000 points (bottom).

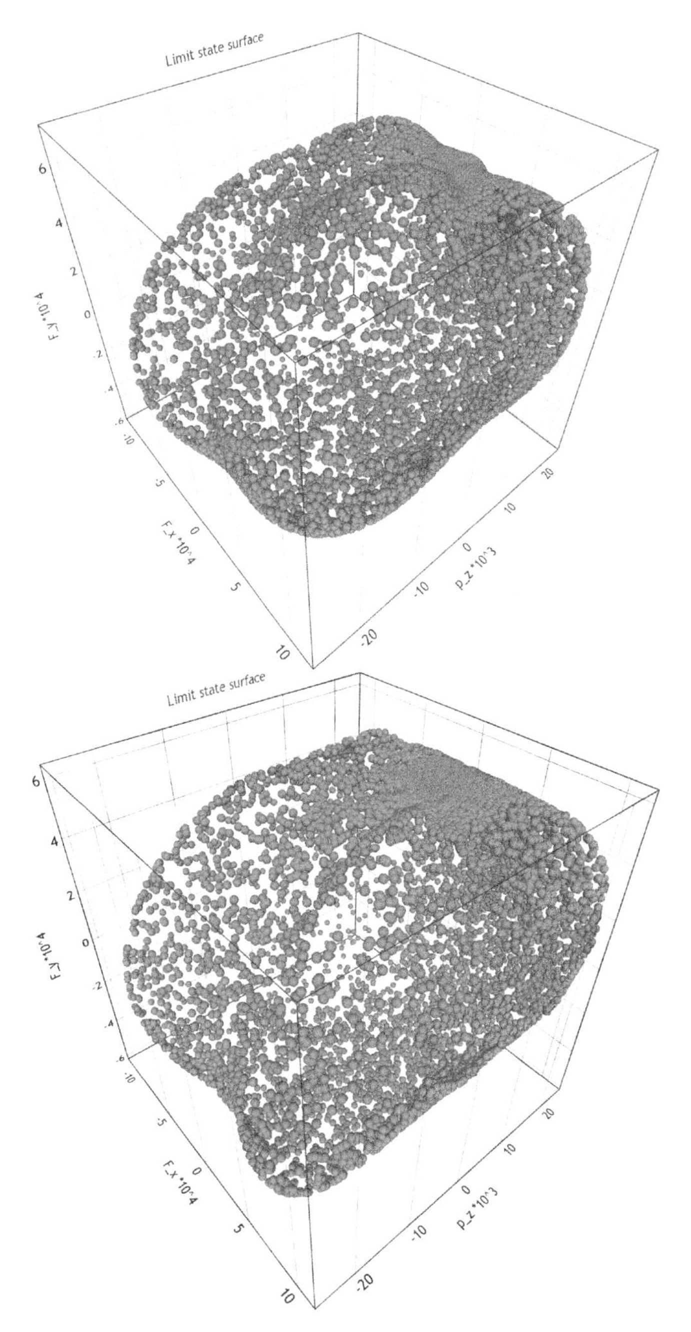

Figure 6.27 Visualization of limit state function $g(\mathbf{x})$, radial basis function interpolation based on 150 points (top) and on 1000 points (bottom).

Table 6.9 Support points for quadratic response surface.

i	p_z [kN/m]	F_x [kN]	F_y [kN]	$g(\mathbf{x}^{(i)})$
1	12.000	30.000	40.000	1
2	21.513	30.000	40.000	0
3	−21.516	30.000	40.000	0
4	12.000	113.180	40.000	0
5	12.000	−81.094	40.000	0
6	12.000	30.000	59.082	0
7	12.000	30.000	−59.087	0
8	19.979	109.790	40.000	0
9	13.527	30.000	59.084	0
10	12.000	45.275	59.094	0

to be $p_f = 7.7 \cdot 10^{-5}$ corresponding to a value of the safety index $\beta = 3.81$. In view of the small number of only 10 support points, this result matches the exact value quite well.

6.4 First passage failure

6.4.1 *Problem formulation*

In dynamic structural analysis, it is usually necessary to apply numerical methods based on a time discretization of the loads and responses (cf. section 4.5). The discretization of the loading process $F(t)$ in time by means of discrete random variables $F_k; k = 1 \ldots n$ at time values t_ℓ typically leads to a large number n of random variables. The response $X(t)$ is then automatically discretized as well, i.e. in terms of its values $X_\ell; \ell = 1, m$ at time points t_ℓ. In the following, we assume $n = m = N$. The first passage problem can then be formulated as the probability that the largest of the response variables becomes larger than a predefined threshold level ξ

$$P_E = \mathbf{Prob}\left[\max_{\ell=1,\ldots m} X_\ell > \xi\right] \tag{6.92}$$

in which it is to be noted that X_ℓ depends on all load variables F_k with a time $t_k \leq t_\ell$, but not those with $t_k > t_\ell$ (principle of causality).

For the sake of notational simplicity, the subsequent derivations will be limited to the case where the vector $\mathbf{F}$ can be represented by one scalar excitation $F(t)$ only. Again, this is an assumption most frequently made in the earthquake analysis of structures

$$\mathbf{F} = \mathbf{P}F(t) \tag{6.93}$$

The process $F(t)$ has a given auto-covariance function $R_{FF}(t_1, t_2)$

$$R_{FF}(t_1, t_2) = E[F(t_1)F(t_2)] \tag{6.94}$$

For quiescent initial conditions, the response of the system (4.64) at any time t can be calculated from Duhamel's integral

$$\mathbf{x}(t) = \int_0^t \mathbf{h}(t-\tau)\mathbf{P}F(\tau)\,d\tau \tag{6.95}$$

in which $\mathbf{h}(u)$ is the impulse response function matrix of the system. One particular component x_i of the vector $\mathbf{x}$ can be calculated from

$$x_i(t) = \sum_{k=1}^{n} \int_0^t h_{ik}(t-\tau)P_k F(\tau)\,d\tau \tag{6.96}$$

From this equation, linear combinations $y(t)$ of the $x_i(t)$ (such as required e.g. for the analysis of internal forces) can be calculated quite easily

$$y(t) = \sum_{i=1}^{M} c_i x_i(t) \tag{6.97}$$

Utilizing a time-discrete representation (frequently called lumped-impulse procedure), the excitation $F(t)$ can be represented by a random pulse train (Lin 1976)

$$F(t) = \sum_{j=1}^{N} F(\tau_j)\,\Delta t\,\delta(t-\tau_j) \tag{6.98}$$

in which Δt is the time step and $\delta(.)$ denotes Dirac's Delta function. Consequently, the response $x_i(t)$ for $t \in [0, T_E]$ can be written as (cf. eq. 6.96)

$$x_i(t) = \sum_{k=1}^{n} \sum_{j=1}^{N} h_{ik}(t-\tau_j)\,P_k\,F(\tau_j)\,\Delta t \tag{6.99}$$

Here, an equidistant spacing with N subdivisions of the interval $[0, T_E]$ is assumed. Obviously, eq. (6.99) represents the response x_i as a linear combination of N Gaussian random variables $F(\tau_j); j = 1 \ldots N$. Based on this observation, it is clear that the probability of reaching or of exceeding the threshold ξ at time t can be directly and accurately calculated by applying the First-Order-Reliability-Method (FORM, Hasofer and Lind 1974). Thus, the limit state condition at time t_r becomes

$$\xi = y(t_r) = \sum_{i=1}^{M} c_i \sum_{k=1}^{N} \sum_{j=1}^{N} h_{ik}(t_r-\tau_j)\,P_k F(\tau_j)\Delta t = \sum_{j=1}^{N} A_{rj} f_j; \quad f_j = F(\tau_j) \tag{6.100}$$

The most probable combination of the f_j's leading to the limit state is then readily calculated by applying FORM. From the auto-covariance function as defined in eq.(6.94) a discrete covariance matrix for the sequence $f_j; j = 1 \ldots N$ can be obtained

$$\mathbf{C_{ff}} = E[\mathbf{f}\mathbf{f}^T] \tag{6.101}$$

which can be Cholesky-decomposed into

$$\mathbf{C}_{\mathbf{ff}} = \mathbf{L}\mathbf{L}^T \tag{6.102}$$

Here, $\mathbf{L}$ is a lower triangular matrix and $\mathbf{f}$ is a vector containing the sequence $f_j; j = 1\ldots N$. The transformation

$$\mathbf{f} = \mathbf{L}\mathbf{u}; \quad \mathbf{u} = \mathbf{L}^{-1}\mathbf{f} \tag{6.103}$$

yields a representation of eq. (6.100) in terms of uncorrelated standardized Gaussian variables u_s

$$\xi = y(t_r) = \sum_{j=1}^{N}\sum_{s=1}^{N} A_{rj}\, L_{js}\, u_s = \sum_{s=1}^{N} b_{rs}\, u_s \tag{6.104}$$

The safety index β_r is easily found from (see e.g. Madsen, Krenk, and Lind 1986)

$$\frac{1}{\beta_r^2} = \sum_{s=1}^{N} \left(\frac{b_{rs}}{\xi}\right)^2 \tag{6.105}$$

The design point $\mathbf{u_r}^*$ (the most likely combination of uncorrelated variables leading to failure at time t_r) is then calculated from

$$u_{rs}^* = \frac{b_{rs}}{\xi}\beta_r^2; \quad s = 1\ldots N \tag{6.106}$$

Due to the linearity of the limit state function, the probability P_{t_r} that the response reaches or exceeds the threshold at time t_r is given by

$$P_{t_r} = \Phi(-\beta_r) \tag{6.107}$$

It should be emphasized that Li and DerKiureghian 1995 obtained similar results for the mean outcrossing rate assuming filtered white noise as input. A numerical study performed by Vijalapura, Conte, and Meghella 2000 gave an analogous result for a nonlinear SDOF-system.

The above results are exploited in order to construct a useful importance sampling scheme. In complete analogy to the importance sampling method for static problems, the above design point excitations can be used as "importance sampling mean" excitations. This mean excitation is simply added to the random excitation process as simulated in the usual way. As there are N possible locations for "design points" (cf. eq. 6.106) it becomes necessary to weight these points appropriately. It is suggested to use the values of $\Phi(-\beta_k)$ as weights (Stix 1983; Macke and Bucher 2003), so that the multi-modal importance sampling density $h_{\mathrm{U}}(\mathbf{u})$ in standard normal space becomes

$$h_{\mathrm{U}}(\mathbf{u}) = \frac{1}{\sum_{r=1}^{N}\Phi(-\beta_r)} \sum_{r=1}^{N} \frac{\Phi(-\beta_r)}{(2\pi)^{N/2}} \exp\left(-\frac{1}{2}\sum_{s=1}^{N}(u_{rs}^* - u_s)^2\right) \tag{6.108}$$

The ratio of original density $f_U(\mathbf{u})$ to importance sampling density $h_U(\mathbf{u})$ for a sample function f_k^m as obtained from the uncorrelated variables u_k^m needs to be calculated based on eq. (6.108). This implies that there will be considerable interaction between the N design points as well as the N limit state conditions.

6.4.2 *Extension to non-linear problems*

The extension to non-linear problems is straightforward. In a first step, the design point excitations have to be determined. This is generally achieved by the application of nonlinear programming tools (Shinozuka 1983). However, a study covering several numerical examples has indicated that in many cases the so-called Rackwitz-Fiessler-Algorithm (Rackwitz and Fiessler 1978) provides a simple and computationally effective method to obtain these design point excitations. For the application of this method it is important to obtain the gradients of the limit-state function in "closed-form". For the special case of the Newmark-solution procedure this has been shown in Li and DerKiureghian 1995.

In the following, this is shown on a somewhat more general basis for systems with arbitrary nonlinearity in the restoring forces. In fact, the calculation of the sensitivities can be reduced to a linear dynamics problem. Starting from the equation of motion of a non-linear system

$$\mathbf{M}\ddot{\mathbf{x}} + \mathbf{C}\dot{\mathbf{x}} + \mathbf{r}(\mathbf{x}) = \mathbf{F}(t) \tag{6.109}$$

it can easily be seen that the partial derivatives of the response vector components $\mathbf{x}(t)$ at time t with respect to the excitation $F(\tau_j)$ at time τ_j are given by

$$\mathbf{M}\frac{\partial \ddot{\mathbf{x}}}{\partial F(\tau_j)} + \mathbf{C}\frac{\partial \dot{\mathbf{x}}}{\partial F(\tau_j)} + \frac{\partial \mathbf{r}(\mathbf{x})}{\partial F(\tau_j)} = \frac{\partial \mathbf{F}(t)}{\partial F(\tau_j)} \tag{6.110}$$

It is convenient to introduce a vector of sensitivities $\mathbf{s}_j$ defined by

$$\mathbf{s}_j = \frac{\partial \mathbf{x}(t)}{\partial F(\tau_j)} \tag{6.111}$$

By using the chain-rule of differentiation

$$\frac{\partial \mathbf{r}(\mathbf{x})}{\partial F(\tau_j)} = \frac{\partial \mathbf{r}(\mathbf{x})}{\partial \mathbf{x}} \frac{\partial \mathbf{x}}{\partial F(\tau_j)} = \mathbf{K}_t \frac{\partial \mathbf{x}}{\partial F(\tau_j)} \tag{6.112}$$

in which $\mathbf{K}_t$ denotes the tangential stiffness matrix of the system at time t, we finally obtain

$$\mathbf{M}\ddot{\mathbf{s}}_j + \mathbf{C}\dot{\mathbf{s}}_j + \mathbf{K}_t\mathbf{s}_j = \mathbf{P}\Delta t\delta(t - \tau_j) \tag{6.113}$$

The last term in Eq. 6.113 arises from the pulse train representation of the excitation as given in Eq. 6.98 together with the assumption made in Eq .6.93. Based on these results, the sensitivities of the limit state $y(t)$, as given by Eq. 6.97, can easily be calculated. Eq. 6.113 shows that the required gradients can be obtained from the impulse response function of a time-variant linear system. The tangential stiffness matrix $\mathbf{K}_t$ of this linear

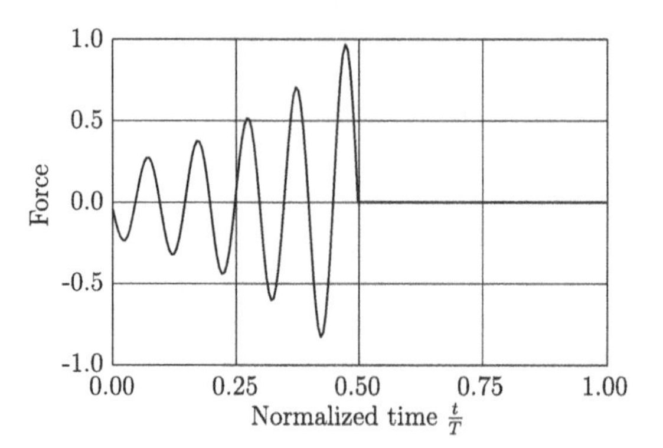

Figure 6.28 Design point excitation for threshold crossing at time $t = 10\pi$ (white excitation).

system is obtained by linearization around the response to the current iterate of the design point excitation. It should be mentioned that for the case of hysteretic systems with internal variables an augmented state-space representation provides a suitable generalization of this approach. However, if the dynamic analysis is to be based on a commercially available FE code, then it will generally not be possible to access all internal variables required. In such a case, numerical gradients may be the last resort. For a detailed review on the topic, see e.g. Macke and Bucher 2003.

Example 6.13 (Linear SDOF-System under stationary white noise excitation)
This example mainly serves the purpose of interpretation of the above mentioned design point excitation. It is a SDOF-system with a natural frequency $\omega_0 = 1$ and a damping ratio of 5% (Naess and Skaug 2000). A time duration $T_E = 20\pi$ is considered. This duration is divided into $N = 200$ time steps, so that $\Delta t = \frac{\pi}{10}$. Let $f(t)$ be a stationary white noise with a two-sided power spectral density $S_0 = \frac{0.1}{\pi}$. Its auto-covariance function is given by

$$R_{ff}(t_1, t_2) = 2\pi S_0 \delta(t_1 - t_2) \tag{6.114}$$

A discrete representation is given in terms of uncorrelated random variables f_i with zero mean and variances $\sigma^2_{f_i} = 2\pi S_0 \Delta t$. Following the above derivations, the design point excitation is given by

$$r_s = h(t - \tau_s)\sqrt{\frac{\Delta t}{2\pi S_0}\frac{\beta^2}{\xi}} \tag{6.115}$$

This is a sequence which is basically a time-reversed impulse response function. For $t = 100\Delta t$ this is shown in Fig. 6.28. The corresponding sample trajectory of $x(t)$ reaches the threshold ξ exactly at time t which is shown in Fig. 6.29. Upon inspection of eq.(6.105) it can easily be seen that in transition to continuous time for white noise excitation we obtain

$$\beta^2 = \frac{\xi^2}{\int_0^t h^2(t - \tau)\, d\tau} = \frac{\xi^2}{\sigma_x^2(t)} \quad \rightarrow \beta = \frac{\xi}{\sigma_x} \tag{6.116}$$

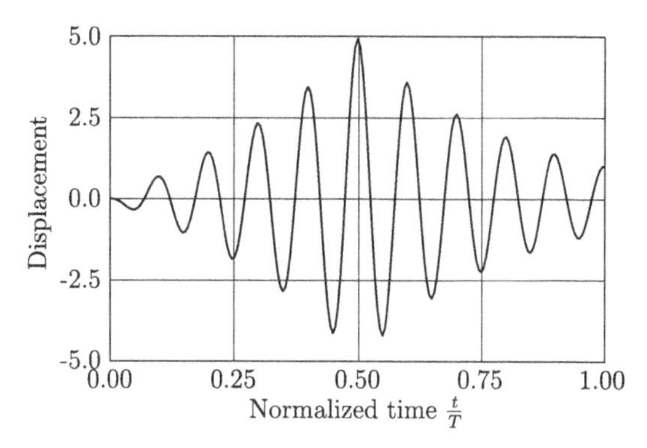

Figure 6.29 Response to design point excitation for threshold crossing at time $t = 10\pi$ (white excitation).

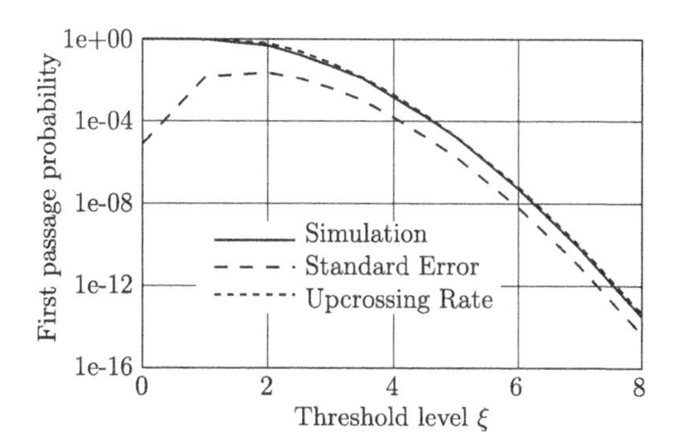

Figure 6.30 First passage probability vs. threshold level for white noise input (500 samples).

The first passage probabilities are evaluated based on the above outlined importance sampling scheme. The threshold level ξ is varied from 0 to $10\sigma_x$. For a sample size of 500 the resulting first passage probabilities are shown in Fig. 6.30. This figure also shows comparative results from an approximation based on the upcrossing rate (Corotis, Vanmarcke, and Cornell 1972). It can be seen that down to the very low probability level of 10^{-15} the coefficient of variation of the estimated probabilities remains at approximately 10%.

Example 6.14 (Non-white excitation) For non-white excitations the picture becomes somewhat more complicated. As an example, consider a stationary excitation with an exponential autocorrelation function (Orenstein-Uhlenbeck process)

$$R_{ff}(t_1, t_2) = \sigma_f^2 \exp\left(-\frac{|t_1 - t_2|}{\tau_c}\right) \tag{6.117}$$

with a correlation time $\tau_c = 2$ and a variance $\sigma_f^2 = 0.02$. This corresponds to a power spectral density $S_{ff}(0) = \frac{\pi}{0.1}$ as in the previous example. In this case, the design point excitation extends into the future as shown in Fig. 6.32 which is a consequence of

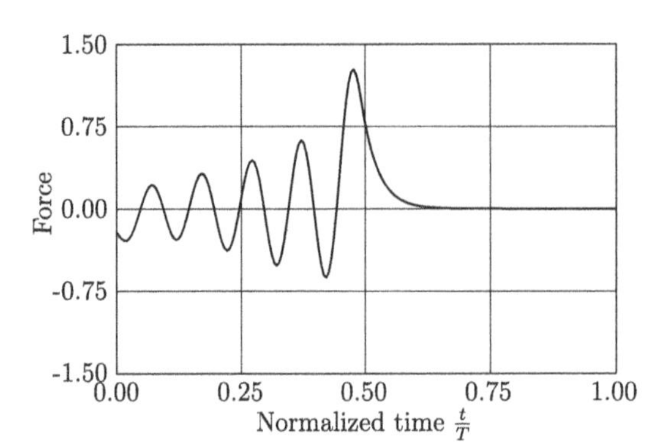

Figure 6.31 Design point excitation for threshold crossing at time $t = 10\pi$ (non-white excitation).

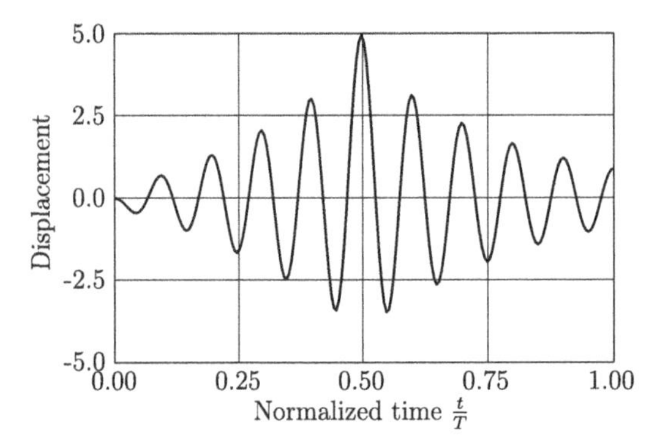

Figure 6.32 Response to design point excitation for threshold crossing at time $t = 10\pi$ (non-white excitation).

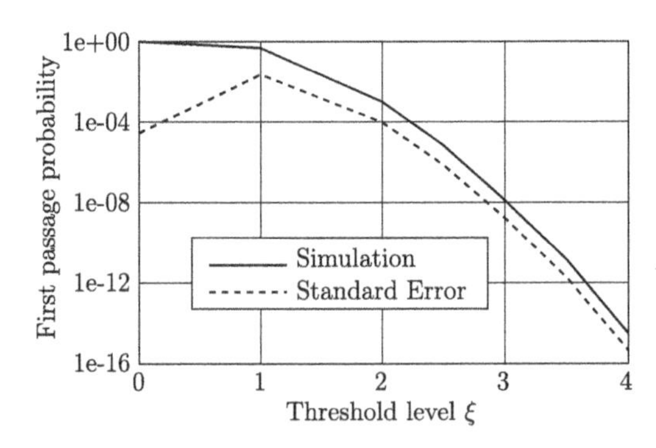

Figure 6.33 First passage probability vs. threshold level for non-white noise input (500 samples).

the temporal correlation. The corresponding response trajectory is given in Fig. 6.32. Again, a time interval $\Delta t = \frac{\pi}{10}$ is chosen for time discretization.

From a sample size of 500 the first passage probabilities as shown in Fig. 6.33 are obtained. Again the c.o.v. remains in the range of 10%, even for probabilities as low as 10^{-15}.

Summary

This chapter started with the definition of the failure probability in terms of a limit state function. First order concepts for the simplified reliability analysis were introduced and applied to various examples. A thorough discussion of Monte-Carlo simulation methods followed. In this context, advanced methods such as importance sampling, adaptive sampling, directional sampling, and asymptotic sampling were extensively investigated. The application of response surface techniques to structural reliability analysis was then presented. Several numerical examples of different complexity were designed to illustrate the potential benefits or shortcomings of these approaches. Finally, the first passage problem of stochastic structural dynamics was re-cast as a reliability problem allowing for the application of the first-order reliability method and importance sampling.

Concluding remarks

The material as presented in this book covers the range from essential basics in probability theory to advanced numerical methods for the treatment of randomness in structural mechanics.

Starting with chapter 2 on probability theory and probabilistic modeling, the required tools for estimation and simulation of random variables were presented. Chapter 3 on regression and response surfaces provided possibilities for approximation relations between random variables in terms of simple functions and methods to eliminate less important variables from the analysis.

Chapter 4 on stochastic structural dynamics discussed the effect of random excitations on the dynamic structural response. Emphasis was put on the treatment of Markov processes which allow for an analytical treatment (e.g. by solving the Fokker-Planck-equation) or a simplified numerical treatment (e.g. by applying covariance analysis). Approximations based on equivalent linearization were discussed. For arbitrarily nonlinear systems, numerical time integration procedures as required for a Monte-Carlo approach were presented. Possible qualitative changes of system behavior due to random excitations was covered in the final section on stochastic stability.

Random fields and stochastic finite elements were discussed in chapter 5. The numerical tools are based on the spectral decomposition of the random fields and their discretization in terms of stochastic finite elements. Semi-analytical approaches based on perturbation theory as well as purely numerical methods based on Monte-Carlo techniques were presented.

Computational methods to determine small probabilities as required for structural reliability analysis were presented in chapter 6. Here the simplifications inherent in first order approximations were discussed in detail. Advanced Monte-Carlo simulation methods suitable for the computation of small probabilities were in the main focus of this chapter. Various methods allowing for sampling the regions of relevance are presented and applied to a variety of structural problems including the first-passage problem in stochastic dynamics.

It is quite obvious that many important topics in stochastic structural analysis could not be covered in this book. One such topic is optimization including randomness such as robust optimization (Bucher 2007). Further steps would lead into the direction of reliability-based optimal structural design and maintenance (e.g. Frangopol and Maute 2003; Bucher and Frangopol 2006; Macke and Higuchi 2007). Another area of importance is system identification and damage detection using non destructive techniques. Here randomness may also play a significant role concerning the quality of the identified results (Bucher, Huth, and Macke 2003).

Nevertheless, it is believed that the contents of this book provides useful tools which can be successfully applied also in the areas not explicitly covered here.

Notations

The following list explains the most frequently used symbols in this book:

$\mathcal{A}, \mathcal{B}, \mathcal{C}$	Events (sets)
$\mathbf{C}_{XX}$	Covariance matrix
D_f	Failure domain
$\mathbf{E}$	Expectation operator
$\mathbf{e}$	Strain tensor in vector notation
$\mathcal{F}$	Failure event
$f_X(x)$	Probability density function
$F_X(x)$	Probability distribution function
$f_{X\vert Y}(x, y)$	Conditional probability density function
$f_{X''}(x)$	Predictive probability density function
$f_{\mathbf{X}}(\mathbf{x})$	Joint probability density function of $\mathbf{X}$
F_e	F-statistic
$\mathbf{F}^e$	Element force vector
$h(t)$	Impulse response function
$H(X)$	Entropy of random variable X
I_e	Coefficient of importance
$\mathbf{K}^e$	Element stiffness matrix
L	Likelihood function
L_c	Correlation length
Prob	Probability
P_E	First passage probability
p_f	Probability of failure
R^2	Coefficient of determination
R^2_{adj}	Adjusted coefficient of determination
R_{XX}	Auto-covariance function
S	Residual sum of squares
S_{XX}	Auto-power spectral density
$\mathbf{s}$	Stress tensor in vector notation
$\mathbf{U}^e$	Element displacement vector
X	Random variable
$\bar{X}$	Expected value of random variable X
σ_X^2	Variance of random variable X
$\mathbf{X}$	Random vector
$\bar{\mathbf{X}}$	Mean value vector

X_k	k-th weighted integral
$\mathbf{x}^{(k)}$	k-th sample point
α	level of significance
β	Reliability index
γ	Euler-Mascheroni constant
$\delta(t)$	Dirac delta function
δW	Virtual work
Γ_m	Estimator from m samples
$\Gamma(x)$	Gamma-function
ζ	Damping ratio
η	Response surface
μ_k	Statistical moment of k-th order
$\boldsymbol{\mu}$	Drift vector
ρ	Coefficient of correlation
$\boldsymbol{\sigma}$	Diffusion vector
ξ	Exceedance threshold
$\varphi(x)$	Standard Gaussian density
$\Phi(x)$	Standard Gaussian distribution
$\Phi^{-1}(x)$	Inverse standard Gaussian distribution
ϕ_k	k-th eigenvector of covariance function
ω_0	Natural circular frequency

Bibliography

Abramowitz, M. and I. A. Stegun (1970). *Handbook of Mathematical Functions.* 9th printing: Dover Publications Inc.

Arnold, L. and P. Imkeller (1994). Fürstenberg-Khasminskii formulas for Lyapunov exponents via anticipative calculus. Technical Report Report Nr. 317, Institut für dynamische Systeme, University of Bremen.

Bathe, K.-J. (1996). *Finite Element Procedures.* Englewood Cliffs: Prentice Hall.

Bjerager, P. (1988). Probability integration by directional simulation. *Journal of Engineering Mechanics 114*, 1285–1302.

Böhm, F. and A. Brückner-Foit (1992). On criteria for accepting a response surface model. *Probabilistic Engineering Mechanics 7*, 183–190.

Box, G. E. P. and N. R. Draper (1959). A basis for the selection of a response surface design. *Journal of the American Statistical Association 54*, 622–654.

Box, G. E. P. and N. R. Draper (1987). *Empirical Model-Building and Response Surfaces.* New York: Wiley.

Box, G. E. P. and K. B. Wilson (1951). On the experimental attainment of optimum conditions. *Journal of the Royal Statistical Society, Series B 13*, 1–45.

Bratley, P. and B. L. Fox (1988). ALGORITHM 659: Implementing Sobol's quasirandom sequence generator. *ACM Transactions on Mathematical Software 14*(1), 88–100.

Breitung, K. W. (1984). Asympotic approximations for multinormal integrals. *Journal of Engineering Mechanics 110*(3), 357–366.

Brenner, C. E. (1995). *Ein Beitrag zur Zuverlässigkeitsanalyse von Strukturen unter Berücksichtigung von Systemuntersuchungen mit Hilfe der Methode der Stochastischen Finite Elemente.* Ph. D. thesis, University of Innsbruck, Austria.

Brenner, C. E. and C. Bucher (1995). A contribution to the SFE-based reliability assessment of nonlinear structures under dynamic loading. *Probabilistic Engineering Mechanics 10*, 265–273.

Bronstein, I. N. and K. A. Semendjajev (1983). *Taschenbuch der Mathematik.* Leipzig: Teubner Verlagsgesellschaft.

Bucher, C. (1988a). Adaptive sampling—an iterative fast Monte Carlo procedure. *Structural Safety 5*(2), 119–126.

Bucher, C. (1988b). Approximate random vibration analysis for MDOF-systems. *Trans. ASME, Journal of Applied Mechanics 55*(1), 197–200.

Bucher, C. (1990). Sample stability of multi-degree-of-freedom systems. In L. Arnold, H. Crauel, and J.-P. Eckmann (Eds.), *Lyapunov Exponents*, pp. 322–330. Springer.

Bucher, C. (1998). Some recent software developments for stochastic structural analysis. In N. Shiraishi, M. Shinozuka, and Y. K. Wen (Eds.), *Structural Safety and Reliability*, pp. 541–547. Rotterdam: Balkema.

Bucher, C. (2001). Stabilization of explicit time imtegration by modal reduction. In W. A. Wall, K.-U. Bletzinger, and K. Schweizerhof (Eds.), *Trends in Computational Structural Mechanics*, pp. 429–437. CIMNE.

Bucher, C. (2006). Applications of random field models in stochastic structural mechanics. In M. Pandey, W.-C. Xie, and L. Xu (Eds.), *Advances in Engineering Structures, Mechanics and Construction*, pp. 471–484. Springer.

Bucher, C. (2007). Basic concepts for robustness evaluation using stochastic analysis. In K.-U. Bletzinger, C. Bucher, F. Duddeck, H. Matthies, and M. Meyer (Eds.), *Efficient Methods for Robust Design and Optimisation – EUROMECH Colloquium 482 (on CD).*

Bucher, C. (2008). Asymptotic sampling for high-dimensional reliability analysis (submitted for review). *Probabilistic Engineering Mechanics.*

Bucher, C. and U. Bourgund (1990). A fast and efficient response surface approach for structural reliability problems. *Structural Safety 7*, 57–66.

Bucher, C. and D. M. Frangopol (2006). Optimization of lifetime maintenance strategies for deteriorating structures considering probabilities of violating safety, condition, and cost thresholds. *Probabilistic Engineering Mechanics* 21(1), 1–8.

Bucher, C., D. Hintze, and D. Roos (2000). Advanced analysis of structural reliability using commercial FE-codes. In E. Oñate, G. Bugeda, and B. Suárez (Eds.), *ECCOMAS Barcelona*, on CD. CIMNE.

Bucher, C., O. Huth, and M. Macke (2003). Accuracy of system identification in the presence of random fields. In A. DerKiureghian, S. Madanat, and J. Pestana (Eds.), *Applications of Statistics and Probability in Civil Engineering*, pp. 427–433. Millpress.

Bucher, C. and M. Macke (2005). Response surface methodology. In E. Nikolaidis, D. M. Ghiocel, and S. Singhal (Eds.), *Structural Reliability Handbook*, pp. 19–1 to 19–23. Boca Raton: CRC Press.

Bucher, C. and T. Most (2008). A comparison of approximate response functions in structural reliability analysis. *Probabilistic Engineering Mechanics 23*, 154–163.

Buhmann, M. D. (2004). *Radial Basis Functions: Theory and Implementations.* Cambridge University Press.

Clough, R. W. and J. Penzien (1993). *Dynamics of Structures* (2nd ed.). New York: McGraw-Hill.

Corotis, R. B., E. H. Vanmarcke, and C. A. Cornell (1972). First passage of nonstationary random processes. In *J. Eng. Mech Div., Proc. ASCE, 98: EM2*, pp. 401–414.

Di Paola, M. and I. Elishakoff (1996). Non-stationary response of linear systems under stochastic gaussian and non-gaussian excitation: a brief overview of recent results. *Chaos, Solitons & Fractals* 7(7), 961–971.

Dimentberg, M. F. (1982). An exact solution to a certain non-linear random vibration problem. *Int. J. Non-Linear Mechanics* 17(4), 231–236.

Ditlevsen, O. (1991). Random field interpolation between point by point measures properties. In *Proceedings of 1. Int. Conference on Computational Stochastic Mechanics*, pp. 801–812. Computational Mechanics Publications.

Ditlevsen, O. and H. O. Madsen (2007). *Structural reliability methods.* Technical University of Denmark, http://www.web.mek.dtu.dk/staff/od/books.htm: Internet edition 2.3.7.

Eaton, J. W. (2008). Octave documentation. http://www.gnu.org/software/octave/doc/interpreter/ (15 December 2008).

Eller, C. (1988). Lineare und nichtlineare Stabilitätsanalyse periodisch erregter visko-elastischer Strukturen. Technical Report 88-2, Institut für konstruktiven Ingenieurbau, Ruhr-University Bochum.

Faravelli, L. (1989). Response-surface approach for reliability analysis. *Journal of Engineering Mechanics 115*, 2763–2781.

Fisher, R. A. (1921). Studies in crop variation. I. An examination of the yield of dressed grain from broadbalk. *J. Agricult. Sci.*, 11, 107–135.

Florian, A. (1992). An efficient sampling scheme: Updated Latin Hypercube Sampling. *Probabilistic Engineering Mechanics 7*, 123–130.

Frangopol, D. M. and K. Maute (2003). Life-cycle reliability-based optimization of civil and aerospace structures. *Computers and Structures 81*, 397–410.

Ghanem, R. and P. D. Spanos (1991). *Stochastic Finite Elements – A Spectral Approach.* New York Berlin Heidelberg: Springer.

Guan, X. L. and R. E. Melchers (1997). Multitangent-plane surface method for reliability calculation. *Journal of Engineering Mechanics 123*, 996–1002.

Hasofer, A. M. and N. C. Lind (1974). Exact and invariant second-moment code format. In *J. Eng. Mech Div., Proc. ASCE, 100: EM1*, pp. 111–121.

Hill, W. J. and W. G. Hunter (1966). A review of response surface methodology: A literature survey. *Technometrics 8*, 571–590.

Hong, H.-S. and F. J. Hickernell (2003). Algorithm 823: Implementing scrambled digital sequences. *ACM Transactions on Mathematical Software 29*(2), 95–109.

Imam, R. L. and W. J. Conover (1982). A distribution-free approach to including rank correlation among input variables. *Communications in Statistics, Part B – Simulation & Computation 11*(3), 311–334.

Itô, K. (2004). *Stochastic Processes – Lectures given at Aarhus University.* Berlin-Heidelberg-New York: Springer.

Khuri, A. I. and J. A. Cornell (1996). *Response Surfaces: Designs and Analyses.* New York: Dekker.

Kim, S.-H. and S.-W. Na (1997). Response surface method using vector projected sampling points. *Structural Safety 19*, 3–19.

Kullback, S. (1997). *Information Theory and Statistics.* Dover.

Lancaster, P. and K. Salkauskas (1981). Surface generated by moving least squares methods. *Mathematics of Computation 37*, 141–158.

Li, C.-C. and A. DerKiureghian (1995). Mean outcrossing rate of nonlinear response to stochastic input. In Lemaire, Favre, and Mébarki (Eds.), *Applications of Statistics and Probability*, pp. 295–302. Rotterdam: Balkema.

Lin, Y. K. (1976). *Probabilistic Theory of Structural Dynamics.* Malabar: Krieger.

Lin, Y. K. and G. Q. Cai (1995). *Probabilistic Structural Dynamics – Advanced Theory and Applications.* New York: McGraw-Hill.

Liu, P.-L. and A. DerKiureghian (1986). Multivariate distribution models with prescribed marginals and covariances. *Probabilistic Engineering Mechanics 1*(2), 105–112.

Macke, M. and C. Bucher (2003). Importance sampling for randomly excited dynamical systems. *Journal of Sound and Vibration 268*, 269–290.

Macke, M. and S. Higuchi (2007). Optimizing maintenance interventions for deteriorating structures using cost-benefit criteria. *Journal of Structural Engineering 133*(7), 925–934.

Madsen, H. O., S. Krenk, and N. C. Lind (1986). *Methods of Structural Safety.* Englewood Cliffs: Prentice-Hall.

Matthies, H. G., C. E. Brenner, C. G. Bucher, and C. G. Soares (1997). Uncertainties in Probabilistic Numerical Analysis of Structures and Solids—Stochastic Finite Elements. *Struct. Safety 19*, 283–336.

Matthies, H. G. and C. Bucher (1999). Finite Elements for Stochastic Media Problems. *Comput. Methods Appl. Mech. Engrg. 168*, 3–17.

Maxima (2008). Maxima documantation. http://maxima.sourceforge.net/docs/manual/en/maxima.html (15 December 2008).

Mead, R. and D. J. Pike (1975). A review of response surface methodology from a biometric viewpoint. *Biometrics 31*, 803–851.

Most, T. and C. Bucher (2005). A moving least squares weighting function for the element-free Galerkin method which almost fulfills essential boundary conditions. *Structural Engineering and Mechanics 21*(3), 315–332.

Most, T., C. Bucher, and Y. Schorling (2004). Dynamic stability analysis of nonlinear structures with geometrical imperfections under random loading. *Journal of Sound and Vibration 1–2*(276), 381–400.

Myers, R. H. (1999). Response surface methodology—current status and future directions. *Journal of Quality Technology 31*, 30–44.

Myers, R. H., A. I. Khuri, and J. W. H. Carter (1989). Response surface methodology: 1966–1988. *Technometrics 31*, 137–157.

Myers, R. H. and D. C. Montgomery (2002). *Response Surface Methodology: Process and Product Optimization Using Designed Experiments*. New York: Wiley.

Nadarajah, S. and S. Kotz (2006). Reliability models based on bivariate exponential distributions. *Probabilistic Engineering Mechanics 21*(4), 338–351.

Naess, A. and C. Skaug (2000). Importance sampling for dynamical systems. In R. E. Melchers and M. G. Stewart (Eds.), *Applications of Statistics and Probability*, pp. 749–755. Rotterdam: Balkema.

Nataf, A. (1962). Détermination des distributions de probabilités dont les marges sont données. *Comptes rendus de l'Académie des sciences 225*, 42–43.

Niederreiter, H. (1992). *Random Number Generation and Quasi-Monte Carlo Methods*. Society for Industrial Mathematics.

Noh, Y., K. Choi, and L. Du (2008). Reliability-based design optimization of problems with correlated input variables using a gaussian copula. *Struct. Multidisc. Optim.* (DOI 10.1007/s00158-008-1277-9).

Ouypornprasert, W., C. Bucher, and G. I. Schuëller (1989). On the application of conditional integration in structural reliability analysis. In A. H.-S. Ang, M. Shinozuka, and G. I. Schuëller (Eds.), *Proc. 5th Int. Conf. on Structural Safety and Reliability, San Francisco*, pp. 1683–1689.

Papoulis, A. (1984). *Probability, Random Variables and Stochastic Processes*. McGraw-Hill.

Rackwitz, R. (1982). Response surfaces in structural reliability. Berichte zur Zuverlässigkeitstheorie der Bauwerke, Heft 67, München.

Rackwitz, R. and B. Fiessler (1978). Structural reliability under combined random load sequences. *Computers and Structures* (9), 489–494.

Roberts, J. B. and P. Spanos (2003). *Random vibration and statistical linearization*. Mineola, New York: Dover Publications Inc.

Roos, D., C. Bucher, and V. Bayer (2000). Polyhedral response surfaces for structural reliability assessment. In R. E. Melchers and M. Stewart (Eds.), *Applications of Statistics and Probability*, pp. 109–115. Balkema.

Schittkowski, K. (1986). NLPQL: A fortran subroutine for solving constrained nonlinear programming problems. *Annals of Operations Research 5*, 485–500.

Schorling, Y. (1997). *Beitrag zur Stabilitätsuntersuchung von Strukturen mit räumlich korrelierten geometrischen Imperfektionen*. Ph. D. thesis, Bauhaus-University Weimar.

Shepard, D. (1968). A two-dimensional interpolation function of irregularly-spaced data. In *Proceedings, 1968 ACM National Conference*, pp. 517–524.

Shinozuka, M. (1983). Basic analysis of structural safety. *Journal of Structural Engineering* (109:3), 721–740.

Sobol, I. and D. Asotsky (2003). One more experiment on estimating high-dimensional integrals by quasi-monte carlo methods. *Mathematics and Computers in Simulation 62*, 255–263.

Stix, R. (1983). *Personal Communication.*

VanMarcke, E. (1983). *Random Fields: Analysis and Synthesis*. Cambridge: MIT Press.

Viertl, R. (2003). Einführung in die Statistik, 3rd ed., Wien New York: Springer.

Vijalapura, P. K., J. P. Conte, and M. Meghella (2000). Time-variant reliability analysis of hysteretic SDOF systems with uncertain parameters and subjected to stochastic loading. In R. E. Melchers and M. G. Stewart (Eds.), *Applications of Statistics and Probability*, pp. 827–834. Rotterdam: Balkema.

Wall, F. J. and C. Bucher (1987). Sensitivity of expected exceedance rate of sdof-system response to statistical uncertainties of loading and system parameters. *Probabilistic Engineering Mechanics 2*(3), 138–146.

Wherry, R. J. (1931). A new formula for predicting the shrinkage of the coefficient of multiple correlation, *Ann. Math. Statist.*, 2(4), 440–457.

Wong, F. S. (1985). Slope reliability and response surface method. *Journal of Geotechnical Engineering 111*, 32–53.

Zheng, Y. and P. K. Das (2000). Improved response surface method and its application to stiffened plate reliability analysis. *Engineering Structures 22*, 544–551.

Subject Index

Structures and Infrastructures Series

Book Series Editor: Dan M. Frangopol

ISSN: 1747–7735

Publisher: CRC/Balkema, Taylor & Francis Group

1. Structural Design Optimization Considering Uncertainties
   Editors: Yiannis Tsompanakis, Nikos D. Lagaros & Manolis Papadrakakis
   2008
   ISBN: 978-0-415-45260-1 (Hb)

2. Computational Structural Dynamics and Earthquake Engineering
   Editors: Manolis Papadrakakis, Dimos C. Charmpis,
   Nikos D. Lagaros & Yiannis Tsompanakis
   2008
   ISBN: 978-0-415-45261-8 (Hb)

3. Computational Analysis of Randomness in Structural Mechanics
   Christian Bucher
   2009
   ISBN: 978-0-415-40354-2 (Hb)

Forthcoming:

4. Frontier Technologies for Infrastructures Engineering
   Editors: Shi-Shuenn Chen & Alfredo H-S. Ang
   2009
   ISBN: 978-0-415-49875-3 (Hb)

For Product Safety Concerns and Information please contact our EU representative GPSR@taylorandfrancis.com Taylor & Francis Verlag GmbH, Kaufingerstraße 24, 80331 München, Germany

Printed and bound by CPI Group (UK) Ltd, Croydon, CR0 4YY
01/05/2025
01858470-0004